W0105939

Progress in Nonlinear Differential Equations and Their Applications
Volume 29

Editor

Haim Brezis
Université Pierre et Marie Curie
Paris
and
Rutgers University
New Brunswick, N.J.

Editorial Board

Antonio Ambrosetti, Scuola Normale Superiore, Pisa
A. Bahri, Rutgers University, New Brunswick
Luis Cafarelli, Institute for Advanced Study, Princeton
Lawrence C. Evans, University of California, Berkeley
Mariano Giaquinta, University of Pisa
David Kinderlehrer, Carnegie-Mellon University, Pittsburgh
Sergiu Klainerman, Princeton University
Robert Kohn, New York University
P. L. Lions, University of Paris IX
Jean Mahwin, Université Catholique de Louvain
Louis Nirenberg, New York University
Lambertus Peletier, University of Leiden
Paul Rabinowitz, University of Wisconsin, Madison
John Toland, University of Bath

Nonlinear Partial Differential Equations in Geometry and Physics

The 1995 Barrett Lectures

Garth Baker
Alexandre Freire
Editors

Springer Basel AG

Garth Baker and Alexandre Freire
Department of Mathematics
University of Tennessee
Knoxville, TN 37996-1300
USA

1991 Mathematics Subject Classification: 81T13, 53C07, 57N13, 35L70

A CIP catalogue record for this book is available from the Library of Congress,
Washington D.C., USA

Deutsche Bibliothek Cataloging-in-Publication Data
Nonlinear partial differential equations in geometry and
physics : the 1995 Barrett lectures / Garth Baker ; Alexandre
Freire ed. – Basel ; Boston ; Berlin : Birkhäuser, 1997
   (Progress in nonlinear differential equations and their applications ; Vol. 29)
   ISBN 978-3-0348-9818-8   ISBN 978-3-0348-8895-0 (eBook)
   DOI 10.1007/978-3-0348-8895-0
NE: Baker, Garth [Hrsg.]; GT

This work is subject to copyright. All rights are reserved, whether the whole or part of the
material is concerned, specifically the rights of translation, reprinting, re-use of illustrations,
broadcasting, reproduction on microfilms or in other ways, and storage in data banks. For any
kind of use whatsoever, permission from the copyright owner must be obtained.

© 1997 Springer Basel AG
Originally published by Birkhäuser Verlag in 1997
Softcover reprint of the hardcover 1st edition 1997
Printed on acid-free paper produced of chlorine-free pulp. TCF ∞

ISBN 978-3-0348-9818-8

9 8 7 6 5 4 3 2 1

Contents

Ronald Fintushel

New Directions in 4-Manifold Theory

Sergiu Klainerman

On the Regularity of Classical Field Theories in Minkowski Space-Time $\mathbb{R}^{3+1}$

Fang-Hua Lin

Static and Moving Vortices in Ginzburg-Landau Theories

Michael Struwe

Wave Maps

Preface

This volume presents the proceedings of a series of lectures hosted by the Mathematics Department of The University of Tennessee, Knoxville, March 22–24, 1995, under the title "Nonlinear Partial Differential Equations in Geometry and Physics".

While the relevance of partial differential equations to problems in differential geometry has been recognized since the early days of the latter subject, the idea that differential equations of differential-geometric origin can be useful in the formulation of physical theories is a much more recent one. Perhaps the earliest emergence of systems of nonlinear partial differential equations having deep geometric and physical importance were the Einstein equations of general relativity (1915). Several basic aspects of the initial value problem for the Einstein equations, such as existence, regularity and stability of solutions remain prime research areas today, eighty years after Einstein's work.

An even more recent development is the realization that structures originally introduced in the context of models in theoretical physics may turn out to have important geometric or topological applications. Perhaps its emergence can be traced back to 1954, with the introduction of a non-abelian version of Maxwell's equations as a model in elementary-particle physics, by the physicists C.N. Yang and R. Mills. The rich geometric structure of the Yang-Mills equations was brought to the attention of mathematicians through work of M.F. Atiyah, N. Hitchin, I. Manin, I.M. Singer and others, dating back to the early 1970's. By the early 1980's, through the work of S.K. Donaldson, C.H. Taubes, K.K. Uhlenbeck and others, studying moduli spaces of anti-self-dual solutions of these systems had led to profound developments in understanding the differential topology of compact simply connected 4-manifolds.

Simultaneously, in the 1970's and 80's, several other researchers pursued the formulation of global geometric and topological questions on manifolds in terms of questions of solvability of systems of nonlinear partial differential equations involving special geometric differential operators, in both Riemannian and Hermitian geometry. Some of the key developments were S.-T. Yau's solution of the Calabi conjecture involving Monge-Ampere equations (1977), Y.-T. Siu's rigidity theorems using harmonic maps (1980), R. Hamilton's work on evolution of Riemannian metrics through curvature-driven flows (1982) and R. Schoen's analysis of the scalar curvature equation, which led to the crucial step in the resolution of Yamabe's conjecture (1984).

The present volume contains four sets of lectures on topics which represent some of the most active areas of research today in the field of partial differential equations originating from problems in geometry, topology or theoretical physics. The lectures are of an expository nature, and directed at graduate students and researchers in the interface of geometric analysis and mathematical physics.

In 1994 N. Seiberg and E. Witten introduced new topological invariants for smooth 4-manifolds, computed from moduli spaces of solutions to equations which have their origin in 4-dimensional "$N = 2$" supersymmetric Yang-Mills theories in quantum chromodynamics. Understanding the relationship between S. Donaldson's theory and these new invariants is a topic of intense current activity. In the first set of lectures, Ronald Fintushel begins with a brief overview of Donaldson's invariants and an introduction to the Seiberg-Witten equations, and describes some of their applications to the theory of differentiable 4-manifolds. His second and third lectures deal primarily with the recent proof of the "Immersed Thom conjecture" using the new invariants, and with very recent progress towards the "11/8 conjecture" – an attempt to characterize the indefinite integral quadratic forms which occur as intersection forms of smooth 4-manifolds.

Sergiu Klainerman's lectures deal with the nonlinear hyperbolic differential equations arising from classical field theories described by a Lagrangean variational principle in Minkowski spacetime. The key issues in the mathematical analysis are break-down of solutions in finite time from smooth initial data, global existence of solutions, and the minimal regularity of the data required for well-posedness of the Cauchy problem. The main examples are wave maps, the Yang-Mills equations and the Einstein field equations. In each case the natural scaling invariance of the equations characterizes the number of space dimensions for which the problem is "critical", "supercritical" or "subcritical". General conjectures regarding local well-posedness and asymptotic behavior are set forth in each case, and the currently available evidence is discussed. Klainerman's lectures then focus on the question of optimal conditions for local well-posedness, including a detailed survey of his recent and ongoing work (jointly with M. Machedon) in which a structural feature of many equations with physical origin ("null forms") plays a key role. This has enabled them to obtain results on local well-posedness for the wave maps and Yang-Mills systems that are much closer to the expected minimal Sobolev regularity class than those known classically.

The topic of Fang-Hua Lin's three survey lectures are the Ginzburg-Landau equations occurring in a classical model for superconductivity. The main mathematical issue discussed is the limiting behavior of vortex lines as the "Ginzburg-Landau parameter" approaches zero, in both static and time-dependent models. The starting point is the recent work of Bethuel, Brezis and Helein on the limiting distribution of vortices for energy minimizers in two-dimensional domains. In these results a key role is played by finite configurations of points which are critical for a "renormalized energy"; the problem of finding Ginzburg-Landau minimizers (or critical points) corresponding to given critical configurations is described in the second lecture. The third lecture deals with the time-dependent problem; specifically, the determination of the precise time scale in which motion of the vortices is observed, and of the finite-dimensional dynamical system describing the motion.

In the final set of lectures, Michael Struwe surveys "wave maps"-harmonic maps from Minkowski spacetime to a Riemannian manifold. Wave maps occur naturally for example in non-linear "sigma models" and in the Einstein vacuum equations in the presence of symmetry. Following a survey of local existence results for the quasilinear hyperbolic system defined by wave maps (obtained by energy methods), Struwe's lectures concentrate on the problems of blow-up, global existence and well-posedness in the energy class (in two space dimensions). The attempt to understand the latter problem has led to weak-compactness results for wave maps obtained by "concentration compactness" techniques, which parallel similar developments in the elliptic and parabolic cases; these are described in the last lecture in the volume.

The 1995 Barrett Lectures received financial support from the National Science Foundation (grant DMS-9404089) and from Science Alliance. The organizational assistance provided by the conference secretary, Ms. Kelly Nicely, was also essential for their success.

Garth A. Baker and Alexandre Freire
April 1996

John H. Barrett (1922–1969)

John H. Barrett was born in 1922 and grew up on a farm In Kansas. He received his A.B. in 1944 from Fort Hays Kansas State College. He did his graduate work at the University of Texas and was an instructor there from 1946–1951. His thesis was on differential equations of non-integer order and under the direction of H.J. Ettlinger. He was an assistant professor at Delaware from 1951–1956, an associate professor at Utah from 1956–1961, and professor at Tennessee from 1961–1969, where he was also head from 1964–1969. He also spent the year 1955–1956 visiting Yale and the year 1959–1960 visiting the Research Center at Wisconsin. In January, 1969, he died from complications following a kidney transplant. His wife, Lida, was also a faculty member at Tennessee, where she was head from 1973 to 1980. They had three children.

John was well known for his work on oscillation and disconjugacy theory of linear differential equations as well as the study of boundary value problems. One of his best known contributions was the extension of the classical Prüfer transformation for 2nd order scalar differential equations to systems of differential equations. He had several Ph.D. students. In a time when a common teaching load was three courses, he regularly ran a seminar on differential equations in addition. Typically it would meet very early on a Tuesday or Thursday morning with John as the most frequent lecturer. Through his efforts and influence, an active group in differential equations evolved at the University of Tennessee. This tradition has continued since that time. The Barrett Lecture series is a fitting reminder of his many contributions to the university and community.

(A tribute delivered by Don Hinton, March 22, 1995)

A List of Previous

John H. Barrett Memorial Lectures

The University of Tennessee, Knoxville
Department of Mathematics

Year	Principal Lecturers
1995	**Ronald Fintushel**
	Michigan State University
	Sergiu Klainerman
	Princeton University
	Fang-Hua Lin
	Courant Institute
	Michael Struwe
	ETH, Zürich
1994	**Robert Gilmer**
	Florida State University
1993	**Donald Dawson**
	Carleton University
	Eugene Dynkin
	Cornell University
	Gopinath Kallianpur
	University of North Carolina, Chapel Hill
1992	**Kyoshi Igusa**
	Brandeis University
1991	**John Ball**
	Herriot-Watt University
1989	**Sir Michael Atiyah**
	Oxford University
	I. M. Singer
	Massachusetts Institute of Technology
	Clifford H. Taubes
	Harvard University
	Karen Uhlenbeck
	University of Texas at Austin
1988	**Alan F. Beardon**
	Cambridge University

1987	**Joyce R. McLaughlin** Rensselaer Polytechnic Institute
1986	**Shreeram S. Abhyankar** Purdue University
1985	**Richard K. Miller** Iowa State University
1984	**Donald Ludwig** University of British Columbia
1983	**David Sarason** University of California, Berkeley
1982	**Jean Mawhin** Catholic University of Louvain
1981	**James H. Bramble** Cornell University
1980	**William T. Eaton** University of Texas at Austin
1979	**Fred Brauer** University of Wisconsin
1978	**Hyman Bass** Columbia University
1977	**W.T. Reid** University of Texas
1976	**Paul Waltman** University of Iowa
1975	**Jack E. Hale** Brown University
1974	**Zeev Nehari** Carnegie-Mellon University
1973	**W.N. Everett** University of Dundee
1972	**Garrett Birkhoff** Harvard University
1970	**Einar Hille,** (Emeritus Yale University) University of New Mexico

List of Participants

Alama, Stan	McMaster Univ.	alama@mcmail.cis.mcmaster.ca
Alexiades, Vasilios	Univ. of Tenn.	vasili@math.utk.edu
Alikakos, Nicholas	Univ. of Tenn.	alikakos@utkvx.utk.edu
Belchev, Evgeni	Michigan State Univ.	belchev@math.msu.edu
Bronsard, Lia	McMaster Univ.	bronsard@icarus.math.mcmaster.ca
Capogna, Luca	Purdue Univ.	capogna@math.purdue.edu
Chmaj, Adam	Univ. of Utah	chmaj@math.utah.edu
Demoulini, Sophia	Univ. of Cal., Davis	demoulin@ucdmath.ucdavis.edu
Deng, Qingping	Univ. of Tenn.	deng@math.utk.edu
Dydak, Jerzy	Univ. of Tenn.	dydak@novell.math.utk.edu
Fintushel, Ron	Michigan State Univ.	ronfint@math.msu.edu
Guidry, Michael	Univ. of Tenn.	guidry@utkvx.utk.edu
Han, Qing	Univ. of Notre Dame	qhan@yansu.math.nd.edu
Hinton, Don	Univ. of Tenn.	hinton@novell.math.utk.edu
Ianakiev, Krassimir	SUNY, Buffalo	ianakiev@newton.math.buffalo.edu
Ishigaki, Makoto	Univ. of Tenn.	
Kepka, Mariusz	Michigan State Univ.	kepka@math.msu.edu
Klainerman, Sergiu	Princeton Univ.	seri@math.princeton.edu
Kowalczyk, Michael	Univ. of Tenn.	
Kukavica, Igor	Univ. of Chicago	kukavica@cs.uchicago.edu
Liang, Min	Univ. of Tenn.	mliang@math.utk.edu
Lin, Fang-Hua	Courant Institute	linf@math.nyu.edu
Lu, Guozhen	Wright State Univ.	gzlu@discover.wright.edu
Mou, Libin	Univ. of Iowa	mou@math.uiowa.edu
del Pino, Manuel	Univ. of Chicago	delpino@math.uchicago.edu
Pinto, Joao	Georgia Tech.	pinto@math.gatech.edu
Plexisakis, Michael	Univ. of Tenn.	plex@math.utk.edu
Promislow, Keith	Penn State Univ.	ksp@math.psu.edu
Schlag, Wilhelm	Univ. of Cal., Berkeley	schlag@math.berkeley.edu
Simpson, Henry	Univ. of Tenn.	simpson@novell.math.utk.edu
Sinkala, Zachariah	Middle Tenn. State Univ.	zsinkala@mtsu.edu
Sowa, Artur	CUNY	sow@cunyvmsl.gc.cuny.edu
Struwe, Michael	E.T.H. Zurich	struwe@math.ethz.ch
Stuart, David	Univ. of Cal., Davis	dmstuart@aztec.ucdavis.edu
Tataru, Daniel	Northwestern Univ.	tataru@math.nwu.edu
Wang, Junping	Brigham Young Univ.	junw@math.byu.edu
Weinstein, Gilbert	Univ. of Alabama, Birmingham	weinstei@vorteb.math.uab.edu
Xiong, Jie	Univ. of Tenn.	xiong@novell.math.utk.edu

Organizers

Baker, Garth	Univ. of Tenn.	garth@math.utk.edu
Freire, Alex	Univ. of Tenn.	freire@math.utk.edu

Progress in Nonlinear Differential Equations
and Their Applications, Vol. 29
© 1997 Birkhäuser Verlag Basel/Switzerland

New Directions in 4-Manifold Theory

RONALD FINTUSHEL*

Department of Mathematics, Michigan State
University East Lansing, Michigan 48824
email: ronfint@math.msu.edu

ABSTRACT. This last fall, Seiberg and Witten introduced exciting new
techniques to the theory of 4-manifolds. My first lecture will serve as an introduction to Donaldson theory and to the Seiberg-Witten equations. I will
outline the major acheivements of both theories and highlight their differ-
ences. Many of the new results that I will speak about in Lecture 1 have
already appeared in print. In my next two lectures I will discuss applications
of Seiberg-Witten Theory which have not yet appeared. In Lecture 2, I will
discuss the invariants of Seiberg and Witten and their use in proving the
'Immersed Thom Conjecture'. In Lecture 3, I will talk about the problem of
realization of intersection forms and the 11/8 Conjecture (still unsolved).

Lecture 1: Donaldson and Seiberg-Witten Invariants

The class of 4-dimensional manifolds holds a unique place in the theory of all
smooth manifolds. Dimension 4 is small enough so that one has a reasonable ex-
pectation for constructing and 'viewing' explicit examples as in lower dimensions.
Furthermore, many examples are provided by algebraic geometry, namely nonsingular complex algebraic surfaces. Also, one might hope to extend to 4 dimensions
some variants of the techniques of surgery which are so successful in higher dimensions. These different points of view made 4-manifold theory a melting pot of
ideas from higher and lower dimensional topology and geometry, and they drove
4-manifold theory until the early 1980's. The article [M] gives a discussion of the
state-of-affairs circa 1980, and one can find a shorter review in [DK, Ch.1].

The goal of this lecture is to outline some of the ideas behind the major
advances in 4-manifold theory that have taken place since 1980. These advances
came in two fronts. The first was led by the work of Simon Donaldson, and the

---

*   The author was partially supported NSF Grant DMS9401032. He would like to
thank Peter Kronheimer for explaining his work on which part of Lecture 3 is
based and Tom Parker for useful comments on an early draft of these notes. He
also wishes to thank the organizers of Barrett Lecture Series for their hospitality.

second by Seiberg and Witten in Fall 1994. Both approaches have very strong ties to mathematical physics and to differential geometry, and both caught the interest and imagination of the general mathematical community.

For simplicity, we shall restrict our lectures to smooth 4-manifolds which are compact and simply connected. This will be assumed, usually without further comment. The most basic invariant of an oriented 4-manifold X is its *intersection form*,

$$Q_X : H_2(X;\mathbf{Z}) \otimes H_2(X;\mathbf{Z}) \to \mathbf{Z}.$$

Each class in $H_2(X;\mathbf{Z})$ is represented by a smoothly embedded oriented surface. If two such surfaces intersect transversely at a point, then ± 1 is assigned to this point according to whether the local orientation obtained from the orientations of the two surfaces agrees or disagrees with the ambient orientation of X. If $a,b \in H_2(X;\mathbf{Z})$, the intersection number $Q_X(a,b)$ which will be standardly denoted by $a \cdot b$ is computed by adding up all the local contributions of a pair of transversely intersecting oriented surfaces which represent a and b. This definition depends only on the homology classes involved, and it defines a nonsingular symmetric bilinear form. It is easy to see, for example, that

$$Q_{\mathbf{CP}^2} = (1) \qquad Q_{S^2 \times S^2} = \begin{pmatrix} 0 & 1 \\ 1 & 0 \end{pmatrix}.$$

According to a theorem of J.H.C. Whitehead [Wh], the homotopy type of a simply connected 4-manifold X is determined by Q_X. It should be mentioned that, even though Q_X seems like such a simple invariant to understand, there remain basic unsolved problems about it. For example, let $b^+(X)$ and $b^-(X)$ be the ranks of the maximum positive definite and negative definite subspaces of Q_X, viewed as a form on $H_2(X;\mathbf{R})$. Then sign $(X) = b^+(X) - b^-(X)$, and the rank of $H_2(X;\mathbf{R})$ is $b_2(X) = b^+(X) + b^-(X)$. A famous conjecture is:

The 11/8 Conjecture. *If X is a simply connected smooth 4-manifold with Q_X even (i.e. with $a \cdot a$ even for all a) then*

$$\frac{b_2(X)}{|\operatorname{sign}(X)|} \geq \frac{11}{8}.$$

Recent advances in studying this conjecture will be the topic of Lecture 3.

In higher dimensions, the problem of studying the diffeomorphism types of manifolds which occur in a given homotopy type is understood by surgery theory. The thrust of 4-manifold theory in the 1970's and early 1980's was to try to understand how surgery theory breaks down in dimension 4. The work of M. Freedman [Fr] shows that in the *topological* category, much of the machinery of surgery theory can be made to work. For example, it follows from Freedman's work that any *smooth* compact simply connected 4-manifolds with isomorphic intersection forms are, in fact, homeomorphic.

The key new idea behind Donaldson's contribution was to take into account the differential geometric structure obtained by imposing a Riemannian metric on a smooth 4-manifold X. To make use of the metric, one considers an auxilliary structure — a vector bundle E over X. Typically in Donaldson theory, this is a complex rank 2 vector bundle with structure group $SU(2)$. Such bundles are classified smoothly by their second chern number $c_2(E)$.

Suppose $c_2(E) = k$. We shall be interested in connections on E. A connection is a differential geometric object which provides the means for taking directional derivatives of sections of E. Given a connection A on E, its curvature is a Lie algebra-valued 2-form $F_A \in \Omega^2(\text{ ad } E)$. (ad E is the bundle of Lie algebras whose fiber over each $x \in X$ is the Lie algebra ad $(E_x) \cong \mathfrak{su}(2)$ where E_x is the fiber of E over the point x.) The curvature F_A measures the obstruction to the commuting of directional derivatives — a connection A with $F_A = 0$ is called flat. The notion of curvature is tied to the topology of E by the Chern-Weil formula

$$c_2(E) = -\frac{1}{8\pi^2} \int_X F_A \wedge F_A.$$

It is important to note that the difference of two connections is a 1-form on X with values in the Lie algebra $\mathfrak{su}(2)$. The space of connections on E (completed in an appropriate Sobolev norm) is denoted by $\mathcal{A}_k$. It is an infinite dimensional affine space, modelled on $\Omega^1(\text{ ad } E)$, the space of 1-forms on X with values in the bundle of Lie algebras associated to E. As such, it is contractible, and so it has little topological content; however the gauge group Aut (E) of bundle automorphisms of E acts on $\mathcal{A}_k$. Its quotient is denoted by $\mathcal{B}_k$. Let $\mathcal{B}_k^*$ denote the open dense subspace of equivalence classes of irreducible connections, i.e. those which do not split as direct sum connections on the direct sum vector bundle $E = L \oplus \bar{L}$ for some complex line bundle L. Then $\mathcal{B}_k^*$ is an infinite dimensional manifold modelled on a Hilbert space.

The homology of $\mathcal{B}_k^*$ is interesting and useful; but it is actually determined by the homotopy type of X. In order to get an object dependent on the smooth structure of X, Donaldson cuts $\mathcal{B}_k^*$ down to a finite-dimensional manifold by imposing the anti-self-duality equation,

$$F_A + *F_A = 0.$$

Here "$*$" is the Hodge star operator determined by the (conformal class of) the metric imposed on X. In terms of a local orthonormal basis for the space of 1-forms, it is given by $*e^{i_1} \wedge e^{i_2} = (-1)^{\text{sgn }(\sigma)} e^{i_3} \wedge e^{i_4}$ where σ is the permutation (i_1, i_2, i_3, i_4) of $(1, 2, 3, 4)$. (It may help to note that $*$ gives the Poincaré duality operator on harmonic forms.) This equation is invariant under the action of Aut (E), and so it actually does descend to $\mathcal{B}_k^*$ where it gives a nonlinear elliptic partial differential equation. The solutions to this equation are called *anti-self-dual* connections, and the set of all their equivalence classes $\mathcal{M}_k \subset \mathcal{B}_k$ forms the moduli space of anti-self-dual connections. The anti-self-duality operator has index $8k - 3(1 + b^+ - b_1)$, and, since we are assuming that X is simply connected, the first betti number $b_1 = 0$. We shall refer to this index $8k - 3(1 + b^+)$ as the "formal dimension" of $\mathcal{M}_k$. The solutions of the anti-self-duality equation satisfy a regularity condition: For a generic Riemannian metric g on X, the moduli space $\mathcal{M}_k(g)$ is a finite dimensional oriented submanifold of $\mathcal{B}_k^*$ provided that $b^+ > 0$ (c.f. [DK]), and its dimension is given by the index $8k - 3(1 + b^+)$.

The left hand side of the anti-self-duality equation $F_A + *F_A = 0$ takes values in matrix-valued 2-forms over X, or more precisely in $\Omega_+^2(\text{ad } E)$. Thus

$$\Omega = \mathcal{A}_k \times_{\text{Aut}(E)} \Omega_+^2(\text{ad } E)$$

is a bundle over $\mathcal{B}_k^*$ whose fibers are these spaces of 2-forms. Then $F_A + *F_A$ may be viewed as a section of this bundle, and $\mathcal{M}_k$ is its zero set. In the finite dimensional analogue, if a section σ of a vector bundle $\Omega \to B$, over a closed oriented manifold, vanishes transversely, then its zero set $Z(\sigma)$ is a smooth submanifold of B and its homology class $[Z(\sigma)]$ is the Poincaré dual of the euler class of the bundle. Thus $Z(\sigma) \in H_*(B; \mathbf{Z})$ depends only on the bundle, not on the choice of section.

In the infinite dimensional case we are describing, the section changes with the metric on X, but Donaldson shows, in like fashion, that certain aspects of $\mathcal{M}_k$ depend only on the smooth structure of X. We could think of $\mathcal{M}_k$ as the (Poincaré dual of the) euler class of Ω. There is trouble in pushing through the analogy with the euler class because $\mathcal{M}_k(g)$ need not be compact. But it is possible to define a compactification so that one has a fundamental class $[\mathcal{M}_k]$.

It is worthwhile to mention the reason why $\mathcal{M}_k(g)$ may fail to be compact. This is due to the Uhlenbeck "bubbling" phenomenon. It occurs when there is a sequence $\{A_n\}$ in $\mathcal{M}_k(g)$ whose curvatures concentrate at a point of X. Namely, a sequence of anti-self-dual connections with $c_2 = k$ may have their curvature concentrate in the vicinity of a point $x \in X$ so that the limit of the curvatures is a multiple of a δ-function at x. The limiting situation corresponds to the disjoint union of an anti-self-dual connection on a bundle over X with $c_2 = j < k$ and a $c_2 = k - j$ anti-self-dual connection on S^4 (an instanton at x).

The Donaldson invariant is defined by means of a canonical homomorphism $\mu : H_i(X; \mathbf{R}) \to H^{4-i}(\mathcal{B}_k^*; \mathbf{R})$. Consider the graded algebra

$$\mathbf{A}(X) = \text{Sym}_*(H_0(X; \mathbf{R}) \oplus H_2(X; \mathbf{R}))$$

where $H_i(X)$ sits in degree $\frac{1}{2}(4 - i)$. The homomorphism μ has an obvious extension: $\mathbf{A}(X) \to H^*(\mathcal{B}_k^*; \mathbf{R})$, and in fact, this gives an isomorphism. The Donaldson invariant D_X is an element of the dual algebra $\mathbf{A}^*(X)$, i.e. a linear function

$$D_X : \mathbf{A}(X) \to \mathbf{R},$$

or equivalently a polynomial on the vector space $H_0(X; \mathbf{R}) \oplus H_2(X; \mathbf{R})$. If $z \in \mathbf{A}(X)$ has degree d and $\dim \mathcal{M}_k = 2d$, then $D_X(z) = \langle \mu(z), [\mathcal{M}_k] \rangle$, the evaluation of the cohomology class $\mu(z)$ on the fundamental class of the (compactified) moduli space. Since the dimension of $\mathcal{M}_k$ is $8k - 3(1 + b^+)$, it follows that D_X is defined only on elements of $\mathbf{A}(X)$ which have degree congruent to $\frac{1}{2}(1 + b^+)$ (mod 4). D_X is defined to be 0 on homogeneous elements of other degrees. Note that D_X captures information from infinitely many bundles since the degree $d = 4c_2(E) - \frac{3}{2}(1 + b^+)$. For simply connected 4-manifolds with $b^+ > 1$ and odd, D_X is a diffeomorphism invariant. (To make this absolutely precise we need to think about orientations, see [DK].)

The Donaldson invariant has led to many significant advances in 4-manifold theory. Two of the most important properties of D_X are the connected sum theorem:

Theorem (DONALDSON [D4]). *If $X = X_1 \# X_2$ is a smooth 4-manifold which is a connected sum of manifolds X_i with $b^+(X_i) > 0$, $i = 1, 2$, then $D_X \equiv 0$.*

and the nonvanishing theorem for algebraic surfaces:

Theorem (DONALDSON [D4]). *Let S be a nonsingular projective algebraic surface. Then $D_S \not\equiv 0$.*

These two theorems provide a method for identifying examples of homeomorphic 4-manifolds which are not diffeomorphic. For example, the simply connected elliptic surface $E(3)$ of geometric genus $p_g = 2$ and without multiple fibers has $Q_{E(3)} = 5(1) \oplus 29(-1)$, and this is exactly the intersection form of $X = 5\mathbf{CP}^2 \# 29\overline{\mathbf{CP}}^2$ where $\overline{\mathbf{CP}}^2$ is $\mathbf{CP}^2$ with the reversed orientation. It follows from the theorems of Whitehead and Freedman that $E(3)$ and X are homeomorphic. However the above theorems of Donaldson show that $D_X \equiv 0$ whereas $D_{E(3)} \not\equiv 0$. Hence the two manifolds are not diffeomorphic. (The first examples of this type were produced by Donaldson in [D2].)

Donaldson invariants have been used to distinguish many other interesting examples. Rather than going more deeply into this aspect of 4-manifold theory, we shall instead recount some of the results of the last two years which have greatly enhanced our understanding of Donaldson theory. If $[1] \in H_0(X; \mathbf{Z}) \cong \mathbf{Z}$ is the generator coming from the orientation, then the corresponding degree 2 element of $\mathbf{A}(X)$ is denoted by 'x'. A 4-manifold is said to have *simple type* if $D_X(x^2 z) = 4D_X(z)$ for all $z \in \mathbf{A}(X)$. Note that this definition involves two different bundles over X since the degree of $x^2 z$ is 4 more than the degree of z. There are currently no known examples of 4-manifolds with $b^+ > 1$ which do not have simple type. If X has simple type, then a construction of Kronheimer and Mrowka [KM1] distills the Donaldson invariant into a single analytic function (or formal power series). Define the *Donaldson series* of X by

$$\mathbf{D}_X : H_2(X; \mathbf{R}) \to \mathbf{R}, \qquad \mathbf{D}_X(\alpha) = D_X((1 + \frac{x}{2}) \exp(\alpha)).$$

The original invariant D_X can be recovered from $\mathbf{D}_X$ by the formula

$$D_X(\alpha^n) = n!\mathbf{D}_X^{(n)}(\alpha)$$

where the superscript '(n)' denotes the homogeneous degree n term. For example, for the $K3$ surface (see Lecture 3) $\mathbf{D}_{K3} = \exp(Q_{K3}/2)$. More generally, we have the following beautiful structure theorem:

Theorem (KRONHEIMER AND MROWKA). *For a simply connected 4-manifold X of simple type there are finitely many basic classes $\kappa_1, \ldots, \kappa_s \in H^2(X; \mathbf{Z})$ such that $\kappa_i \equiv w_2(X) \pmod 2$ (κ_i is characteristic) for each i, and rational numbers $\{a_i\}$ such that*

$$\mathbf{D}_X = \exp(Q_X/2) \sum_{1=1}^{s} a_i e^{\kappa_i}$$

as analytic functions. Furthermore, for each surface Σ smoothly embedded in X and representing a homology class σ such that $\sigma \cdot \sigma \geq 0$ we have the following inequality for the genus $g(\Sigma)$:

$$2g(\Sigma) - 2 \geq \sigma \cdot \sigma + |\kappa_i \cdot \sigma|$$

for all i.

Note that this theorem tells us that even though the Donaldson invariant D_X needs infinitely many bundles for its definition, for manifolds of simple type, D_X is prescribed by a finite number of characteristic cohomology classes and rational numbers. This is proved by a study of singular connections. For a different approach to the structure theorem see [FS3]. The inequality in the above theorem is called the 'adjunction inequality'. This terminology comes from the adjunction formula for complex manifolds. In the case of a complex surface S with canonical class K_S, if Σ is a nonsingular complex curve in S representing σ, then $2g(\Sigma) - 2 = \sigma \cdot \sigma + K_S \cdot \sigma$. There is a similar adjunction formula proved in [FS3]. There one works with immersed 2-spheres and their number p of positive double points rather than with embedded surfaces, but one is able to drop the assumption that $\sigma \cdot \sigma \geq 0$.

Theorem (FINTUSHEL AND STERN [FS3]). *Let X be a simply connected 4-manifold of simple type, and let κ_i be the basic classes. Suppose that the homology class σ is represented by an immersed 2-sphere with p positive (and an arbitrary number of negative) double points. Then*

$$2p - 2 \geq \sigma \cdot \sigma + |\kappa_i \cdot \sigma|$$

for all i, unless $p = 0$ and $\kappa_i \pm 2\sigma$ is also basic, in which case

$$0 \geq \sigma \cdot \sigma + |\kappa_i \cdot \sigma|.$$

It follows that the set $\{\kappa_i\}$ of basic classes of X is a diffeomorphism invariant of X, and the adjunction inequalities indicate that the basic classes serve as smooth replacements for the canonical class of an algebraic surface.

One of the most useful constructions in 4-manifold theory is the operation of blowing up. This is the smooth analogue of the usual complex analytic construction and is simply the process of taking the connected sum with $\overline{\mathbf{CP}}^2$. The homology $H_2(\overline{\mathbf{CP}}^2; \mathbf{Z}) \cong \mathbf{Z}$ is generated by the class E of the exceptional curve, i.e. of $\overline{\mathbf{CP}}^1 \subset \overline{\mathbf{CP}}^2$. One can often simplify a problem by blowing up (we will see an example of this in Lecture 2), and understanding the invariants of a blown up manifold plays a key role in the proof of the structure theorem. Thus it is necessary to determine the effect of blowing up on the Donaldson invariant.

Theorem (FINTUSHEL AND STERN [FS2]). *As functions on* $\mathbf{A}(X)$:

$$D_{X \# \overline{\mathbf{CP}}^2}(E^k) = D_X(B_k(x))$$

where $B_k(x)$ are polynomials (the blowup polynomials) in the degree 2 class $x \in$ $\mathbf{A}(X)$ and are given by a generating function

$$B(x,t) = \sum B_k(x)\frac{t^k}{k!} = e^{\frac{-t^2 x}{6}}\sigma_3(x,t)$$

where $\sigma_3(x,t)$ is a quasi-periodic Weierstrass sigma-function associated to the elliptic curve

$$z^2 = 4y^3 - 4(\frac{x^2}{3} - 1)y - \frac{8x^3 - 36x}{27}.$$

To conclude our brief discussion of Donaldson theory, we mention the problem of actually calculating the Donaldson invariant of a given 4-manifold X. Donaldson showed in [D4] that for algebraic surfaces there are algebro-geometric techniques for calculating D_X. Such calculations are usually very long and intricate. In [FS4], Stern and I introduced an operation which we call 'rational blowdown' which is an analogue of blowing down an exceptional curve in a complex surface. Instead of blowing down a single 2-sphere, rationally blowing down removes a configuration of embedded 2-spheres. There are two key facts about this operation: first, many interesting 4-manifolds can be constructed from known 4-manifolds by this construction, and second, its effect on the Donaldson series is computable.

Here is a sample application. Simply connected elliptic surfaces are classified up to deformation-type by three positive integers, the euler number $e = 12n$, and the multiplicities of the (possible) multiple fibers p and q, which are relatively prime. (The case of fewer than 2 multiple fibers is indicated by p or q being equal to 1. There are elliptic surfaces with more than 2 multiple fibers, but these cannot be simply connected.)

Theorem (FINTUSHEL AND STERN [FS4]). *Let* $E(n; p, q)$ *be a simply connected elliptic surface, and denote by* f *the homology class of a generic fiber. Then*

$$\mathbf{D}_{E(n;p,q)} = \exp(Q/2)\frac{\sinh^n(f)}{\sinh(f/p)\sinh(f/q)}.$$

We now turn to the second major advance in smooth 4-manifold theory, due to N. Seiberg and E. Witten. Although Seiberg-Witten theory was discovered using techniques of physics (see [SW1, SW2, W]), one can take the point of view that its purpose is the characterization of the basic classes of X. Recall that the basic classes $\kappa_i \in H^2(X; \mathbf{Z})$ are characteristic ($\kappa_i \equiv w_2(X)$ (mod 2)). This means that each κ_i gives rise to a spin ^{c}-structure on X. The group $Spin^c(4) \cong (U(1) \times SU(2) \times SU(2))/\{\pm 1\}$ which fibers over $SO(4) \cong (SU(2) \times SU(2))/\{\pm 1\}$ with fiber $S^1 \cong U(1)$. The choice of a Riemannian metric on X gives a reduction of the structure group of the bundle F_X of tangent frames to $SO(4)$, and a spin ^{c}-structure on X is a lift of F_X to a principal $Spin^c(4)$ bundle $\tilde{F}_X$ over X. Since $U(2) \cong (U(1) \times SU(2))/\{\pm 1\}$, we get representations $s^{\pm} : Spin^c(4) \to U(2)$, and associated rank 2 complex vector bundles

$$W^{\pm} = \tilde{F}_X \times_{s^{\pm}} \mathbf{C}^2$$

called spinor bundles.

If X is a spin manifold ($w_2(X) = 0$) then there are spinor bundles $V^{\pm}$ with fiber $\mathbf{C}^2$ constructed from a principal $Spin(4)$ bundle which is a double cover of F_X. If $w_2(X)$ is nonzero, the bundles $V^{\pm}$ do not exist. Think of principal G-bundles as being given by transition functions $\varphi_{ij} : U_i \cap U_j \to G$ which are used to glue together local trivializations to form the given bundle. These functions are defined on the intersections of open sets in some cover for X which can be chosen so that the overlaps $U_i \cap U_j$ are contractible. The transition functions of F_X are maps to $SO(4)$. Since $\pi_1(U_i \cap U_j) = 0$ these maps can each be lifted to $Spin(4) \cong SU(2) \times SU(2)$. The problem is that if $w_2(X) \neq 0$, these lifts can never be chosen to be compatible i.e. they cannot be chosen to satisfy the cocycle condition $\varphi_{ij}\varphi_{jk} = \varphi_{ik}$. However, there is another problem with the same $\mathbf{Z}_2$ obstruction. Consider trying to find the square root of a complex line bundle L whose chern class $c_1(L)$ is congruent to $w_2(X)$ mod 2. This can be accomplished provided $c_1(L)$ is divisible by 2; in other words, if $w_2(X) = 0$. If the transition functions of L are g_{ij} then transition functions for the square root of L are given locally by $g_{ij}^{1/2}$. Again, these cannot be chosen compatibly unless $w_2(X) = 0$. This means that the obstruction to constructing the bundles $W^{\pm} = V^{\pm} \otimes L^{1/2}$ is $2w_2(X) = 0 \in H^2(X; \mathbf{Z}_2)$.

Each orientable 4-manifold admits complex line bundles whose first chern class is characteristic. Hence each orientable 4-manifold admits a spin ^{c}-structure [HH]. The bundle L above is the determinant $\det(W^{\pm})$. For manifolds X without

2-torsion in their first homology, spin $^{\mathbb{C}}$-structures on X are in 1-1 correspondence with the characteristic complex line bundles L over X. As for ordinary spin-structures, one has Clifford multiplication

$$c : T^*X \otimes W^\pm \to W^\mp$$

written $c(v, w) = v.w$ and satisfying $v.(v.w) = -|v|^2 w$. Thus c induces a map

$$c : T^*X \to \text{Hom } (W^+, W^-).$$

A connection A on L together with the Levi-Civita connection on the tangent bundle of X form a connection $\nabla_A : \Gamma(W^+) \to \Gamma(T^*X \otimes W^+)$ on W^+. This connection, followed by Clifford multiplication, induces the Dirac operator

$$D_A : \Gamma(W^+) \to \Gamma(W^-).$$

Thus D_A depends both on the connection A and the Riemannian metric on X. The case where $L = \det(W^+)$ is trivial corresponds to a usual spin structure on X, and in this case we may choose A to be the trivial connection and then $D_A = D : V^+ \to V^-$, the usual Dirac operator.

Given a spin $^{\mathbb{C}}$-structure on X, the bundle of self-dual 2-forms Ω_X^+ of X is also associated to $\tilde{F}_X$ by $\Omega_X^+ \cong \tilde{F}_X \times_{SU(2)} \mathfrak{su}(2)$ where $SU(2)$ acts on its Lie algebra $\mathfrak{su}(2) \cong \mathbb{C} \oplus \mathbb{R}$ via the adjoint action. The map

$$\mathbb{C} \oplus \mathbb{C} \to \mathbb{C} \oplus \mathbb{R} \qquad (z, w) \to (z\bar{w}, |z|^2 - |w|^2)$$

is $SU(2)$ equivariant, and so it induces a map

$$q : \Gamma(W^+) \to \Omega_X^+.$$

Given a pair $(A, \psi) \in \mathcal{A}_X(L) \times \Gamma(W^+)$, i.e. A a connection in $L = \det(W^\pm)$ and ψ a section of W^+, the monopole equations of Seiberg and Witten [W] are

$$\begin{aligned} D_A \psi &= 0 \\ F_A^+ &= q(\psi) \end{aligned}$$

The gauge group Aut $(L) = $ Map (X, S^1) acts on the space of solutions to these equations via $g \cdot (A, \psi) = (A - g^{-1}dg, g\psi)$ and its orbit space is the *Seiberg-Witten moduli space* $M_L(X)$.

Some important features of the Seiberg-Witten equations are:

(1) If (A, ψ) is a solution to the Seiberg-Witten equations with $\psi \neq 0$ then its stabilizer in Aut (L) is trivial. Such solutions are called *irreducible*. The stabilizer of a *reducible* solution $(A, 0)$ consists of the constant maps in Map (X, S^1). This is a copy of S^1.

(2) $(A, 0)$ is a reducible solution if and only if A is an anti-self-dual connection on the complex line bundle L. If $b_X^+ > 0$ and $c_1(L) \neq 0$, a generic metric on X admits no such connections.

(3) The formal dimension of the Seiberg-Witten moduli space is calculated by the Atiyah-Singer theorem to be

$$\dim M_L(X) = \frac{1}{4}(c_1(L)^2 - (3\operatorname{sign}(X) + 2e(X))$$

where $e(X)$ is the euler number of X. Especially interesting is the case where $\dim M_L(X) = 0$, since this is precisely the condition for X to admit an almost-complex structure with first chern class equal to $c_1(L)$.

(4) A self-dual 2-form η on X gives us a perturbation of the Seiberg-Witten equations:

$$\begin{aligned} D_A\psi &= 0 \\ F_A^+ &= q(\psi) + \eta, \end{aligned}$$

and for a generic such perturbation η, the corresponding moduli space of solutions $M_X(L, \eta)$ will be an oriented manifold whose dimension is $\dim M_L(X)$, provided $M_X(L, \eta)$ contains at least one irreducible solution. (As in (2), if $b^+(X) > 0$ and $c_1(L) \neq 0$, all solutions will be irreducible for a generic choice of metric or perturbation η.) For simplicity we shall ignore this perturbation and write $M_L(X)$ for $M_X(L, \eta)$.

(5) There is a Lichnerowicz-type theorem, proved, as usual, with an application of the Weitzenbock formula [W, KM2]: If X carries a metric of positive scalar curvature, then the only solutions of the Seiberg-Witten equations are reducible (of the form $(A, 0)$). Hence, if $b_X^+ > 0$, for a generic metric of positive scalar curvature, $M_L(X) = \emptyset$.

(6) For each L, the Seiberg-Witten moduli space $M_L(X)$ is compact.

(7) There are only finitely many characteristic line bundles L on X for which both $M_L(X) \neq \emptyset$ and $\dim M_L(X) \geq 0$.

Items (6) and (7) are also proved using the Weitzenbock formula in [W, KM2]. Witten has given an interesting intuitive explanation for the failure of bubbling to occur: If a sequence of solutions limits to a bubble, it would mean that there is produced on the round 4-sphere a nontrivial solution (A, ψ) to the Seiberg-Witten equations. Since the round 4-sphere has positive scalar curvature, $\psi = 0$, hence A is anti-self-dual. But the only complex line bundle over S^4 is trivial, and the only anti-self-dual connection on that bundle is also trivial. Thus bubbling cannot occur.

In case dim $M_L(X) = 0$, items (4) and (6) imply that that generically, $M_L(X)$ is a finite set of signed points. In this case one defines the *Seiberg-Witten invariant* $SW_X(L)$ to be the signed count of these points. This invariant can also be defined if dim $M_L(X) \neq 0$. We leave this to Lecture 2, where we also explain why it is a diffeomorphism invariant provided $b^+ > 1$. In contrast to the Donaldson invariant, the Seiberg-Witten invariant is often easily computed. For example, Witten [W] has given the complete calculation of the invariants of Kähler surfaces. In particular, for the canonical class K_X the invariant $SW_X(K_X) \neq 0$. Furthermore, the rational blowdown formulas of [FS4] extend to the Seiberg-Witten invariants with much easier proofs.

Shortly after the discovery of the Seiberg-Witten invariants, proofs of some long-standing conjectures and remarkable new theorems followed:

Theorem (KRONHEIMER AND MROWKA [KM2]). *For a smoothly embedded surface Σ in $\mathbf{CP}^2$ representing d times the generator H of $H_2(\mathbf{CP}^2; \mathbf{Z})$, the genus of Σ must be at least the genus of a nonsingular complex curve of degree d. I.e.*

$$g(\Sigma) \geq \frac{(d-1)(d-2)}{2}.$$

The next important result can be attributed to Friedman and Morgan, Kronheimer and Mrowka, Witten, and perhaps others.

Theorem. *The canonical class K_S of a minimal Kähler surface S is (up to sign) a diffeomorphism invariant.*

Cliff Taubes has used the Seiberg-Witten invariants to study symplectic 4-manifolds. It is a consequence of the following theorem of his that connected sums of 4-manifolds with $b^+ \geq 1$ do not admit symplectic forms compatible with the given orientations.

Theorem (TAUBES [T]). *Let (X, ω) be a symplectic 4-manifold, oriented by $\omega \wedge \omega$. Then if $c_1(L)$ is c_1 of the associated almost complex structure, $SW_X(L) = \pm 1$.*

Let us illustrate some arguments using Seiberg-Witten theory by proving a version of Donaldson's connected sum theorem (cf. [W]). This theorem states that if $b^+(X_1)$ and $b^+(X_2)$ are positive, then the Seiberg-Witten invariants of the connected sum $X = X_1 \# X_2$ vanish. Let L be a characteristic complex line bundle over X which comes from the connected sum of line bundles L_i over X_i. Counting formal dimensions using the formula in (3), we see that dim $M_X(L) =$ dim $M_{X_1}(L_1) + $ dim $M_{X_2}(L_2) + 1$. Since we wish to discuss $SW_X(L)$, we assume that dim $M_X(L) = 0$. Thus one of the dim $M_{X_i}(L_i) < 0$, say for $i = 1$. Thus by (4), $M_{X_1}(L_1) = \emptyset$ provided $c_1(L_1) \neq 0$. (We are assuming $b^+(X_1) > 0$.) This can always be arranged by a trick: First a very simple version of the blowup formula holds for Seiberg-Witten invariants, namely, $SW_{X \# \overline{\mathbf{CP}}^2}(L \pm E) = SW_X(L)$. (A more general version of this is described in [FS5].) Let $\bar{X} = (X_1 \# \overline{\mathbf{CP}}^2) \# (X_2 \# \overline{\mathbf{CP}}^2)$. Then

$SW_{\bar{X}}(L+E_1+E_2) = SW_X(L)$, and neither $c_1(L_1+E_1)$ nor $c_1(L_2+E_2)$ is 0. Thus we may simply assume that $M_{X_1}(L_1) = \emptyset$. Now $X = (X_1 \setminus B^4) \cup S^3 \times [-r, r] \cup (X_2 \setminus B^4)$. Suppose we consider a sequence of metrics on X for which the neck length $2r$ goes to ∞. Seiberg-Witten solutions on $S^3 \times \mathbf{R}$ are asymptotically trivial due to the positive scalar curvature of S^3, and one can argue from this that all solutions of the Seiberg-Witten equations on X for a metric with large enough neck length must be obtained by joining solutions on X_1 to solutions on X_2 (and also that any (A, ψ) obtained this way can be perturbed to a solution). Since there are generically no solutions on X_1, it follows that $SW_X(L) = 0$.

It is worth remarking where the extra "+1" comes from in the dimension formula for the connected sum. Let us drop the assumption that $\dim M_X(L) = 0$ and use the notation $(A_1, \psi_1) \# (A_2, \psi_2)$ for a solution in $M_X(L)$ that comes from irreducible solutions on X_1 and X_2. Let $g \in$ Map (X_2, S^1) vary over the constant maps and consider the solutions $\{\alpha_g = (A_1, \psi_1) \# g \cdot (A_2, \psi_2)\} \in M_X(L)$. (These are all solutions because $g \cdot (A_2, \psi_2)$ is equivalent to (A_2, ψ_2) in $M_{X_2}(L_2)$.) Now $\alpha_{g'} = \alpha_{g''} \in M_X(L)$ if and only if there is an $f = f_1 \# f_2 \in$ Map (X, S^1) such that $f_1 \cdot (A_1, \psi_1) = (A_1, \psi_1)$ and $f_2 g'(A_2, \psi_2) = g''(A_2, \psi_2)$. Since we are assuming that both the (A_i, ψ_i) are irreducible solutions, the stabilizers of (A_i, ψ_i) are trivial; so $f_1 = 1$ and $f_2 g' = g''$. Thus $f = 1 \# g''(g')^{-1}$, but g' and g'' are constant maps; hence we see that $f = 1$. This gives a whole circle of inequivalent elements of $M_X(L)$. In fact, this describes a free S^1-action on $M_X(L)$ provided neither $M_{X_1}(L_1)$ nor $M_{X_2}(L_2)$ contains reducible solutions. If $b^+(X_1)$ and $b^+(X_2)$ are both positive, this S^1 action will imply the triviality on $X = X_1 \# X_2$ of the more general Seiberg-Witten invariants which we shall define in Lecture 2.

We conclude this lecture with a conjecture of Witten about the relationship between Donaldson and Seiberg-Witten invariants. Recall that the structure theorem of Kronheimer and Mrowka states that for a manifold of simple type, the Donaldson series, $\mathbf{D}_X$, is given by

$$\mathbf{D}_X = \exp(Q_X/2) \sum_{1=1}^{s} a_i e^{\kappa_i}$$

where the basic classes κ_i are characteristic.

Witten's Conjecture. X *has simple type if and only if all characteristic line bundles* L *on* X *with* $SW_X(L) \neq 0$ *have* $\dim M_X(L) = 0$. *(There are only finitely many of these by item (7) above.) Then the basic classes are exactly* $c_1(L)$ *for these bundles. Furthermore,*

$$\mathbf{D}_X = 2^{3 \text{ sign } +2} \, \mathrm{e}^{-(\frac{b^+-3}{2})} \exp(Q_X/2) \sum SW_X(L) e^{c_1(L)}.$$

This conjecture is currently known to hold for all elliptic surfaces and all blowups, blowdowns, and rational blowdowns of manifolds which satisfy the conjecture [FS4].

Lecture 2: The Immersed Thom Conjecture

As we mentioned in our first lecture, the classical Thom Conjecture states that the genus of a smoothly embedded surface F in $\mathbf{CP}^2$ representing d times the generator H of $H_2(\mathbf{CP}^2; \mathbf{Z})$ must be at least the genus of a nonsingular complex curve of degree d, i.e.

$$g(F) \geq \frac{(d-1)(d-2)}{2}$$

This was proved by Peter Kronheimer and Tom Mrowka using Seiberg-Witten theory [KM2]. Instead of thinking about representing 2-dimensional homology classes by embedded surfaces, one often wishes to represent them by immersed 2-spheres. This leads to what we call the Immersed Thom Conjecture. In this lecture we shall discuss the following theorem.

Theorem (The Immersed Thom Conjecture) *Suppose that a 2-sphere S is transversely immersed in $\mathbf{CP}^2$ with p positive (and say n negative) double points, and suppose that its image represents $dH \in H_2(\mathbf{CP}^2; \mathbf{Z})$. Then*

$$p \geq \frac{(d-1)(d-2)}{2}.$$

The proof that I will outline is due to Ron Stern and myself [FS5]. Tom Mrowka also has a proof of this theorem. The classical Thom Conjecture directly implies the immersed Thom Conjecture when $n = 0$, because one can remove double points by adding handles, at the expense of raising the genus by one for each double point.

Our proof uses Seiberg-Witten invariants; so let us first review their construction. Given a (simply connected) 4-manifold X with b^+ odd, and a characteristic line bundle L on X, the corresponding Seiberg-Witten moduli space M_L has formal dimension

$$\begin{aligned}
\dim M_L &= \tfrac{1}{4}(c_1(L)^2 - (3 \text{ sign} + 2 \text{ e })(X)) \\
&= \tfrac{c_1(L)^2 - \text{sign} (X)}{4} - (1 + b^+) = 2n.
\end{aligned} \tag{1}$$

This is even because we have assumed that b^+ is odd and because in an inner product space over Z, the square of a characteristic element is congruent to the signature mod 8. If $\dim M_L = 0$ and M_L contains no reducible solutions (solutions of the form $(A, 0)$ where A is an anti-self-dual connection on L), then M_L is a compact oriented 0-dimensional manifold, and $SW_X(L)$ is the count of these points with signs. If $\dim M_L > 0$ then we consider the basepoint map

$$\tilde{M}_L = \{ \text{ solutions } (A, \psi)\}/ \text{ Aut }^0(L) \to M_L$$

where $\text{Aut }^0(L)$ consists of gauge transformations which are the identity on the fiber of L over a fixed basepoint in X. If there are no reducible solutions, the basepoint map is an S^1 fibration, and we denote its euler class by $\beta \in H^2(M_L; \mathbf{Z})$.

Now M_L represents an integral cycle in the configuration space $B_L = (\mathcal{A}_L \times W^+, L)/ \operatorname{Aut}(L)$, and if $\dim M_L = 2n$, the Seiberg-Witten invariant is defined to be the integer

$$SW_X(L) = \langle \beta^n, [M_L] \rangle.$$

Note that the space $\mathcal{A}_L \times \Gamma(W^+)$ is contractible and $\operatorname{Aut}(L) \cong \operatorname{Map}(X, S^1)$ which acts freely on $\mathcal{A}_L \times (\Gamma(W^+) \backslash \{0\})$. Since S^1 is a $K(\mathbf{Z}, 1)$ and we are assuming that X is simply connected, the quotient

$$B_L^* = \left(\mathcal{A}_L \times (\Gamma(W^+) \backslash \{0\}) \right) / S^1$$

of this action is homotopy equivalent to $\mathbf{CP}^\infty$. So if there are no reducible solutions, we may view $M_L \subset \mathbf{CP}^\infty$. Under these identifications, the class β becomes the standard generator of $H^2(\mathbf{CP}^\infty; \mathbf{Z})$.

If $b^+(X) \geq 2$, the map

$$SW_X : \{ \text{ characteristic line bundles on } X \} \to \mathbf{Z}$$

is a diffeomorphism invariant. To see this we must understand why $SW_X(L)$ does not depend on the (generic) choice of Riemannian metric on X used to write down the Seiberg-Witten equations. Taking into account the metric g, write $M_L(g)$ for the corresponding moduli space. We let $\mathcal{C}$ denote the (connected) space of metrics on X, and as in Donaldson theory (cf. [DK]) we consider the parametrized moduli space

$$\mathbf{M}_L = \{(A, \psi, g) | (A, \psi) \in M_L(g)\} \subset B_L \times \mathcal{C}$$

We have a Fredholm projection

$$\pi : \mathbf{M}_L^* = \mathbf{M}_L \backslash \{ \text{ reducible solutions } \} \to \mathcal{C}$$

and, in fact, by the Sard-Smale Theorem the regular values of π are generic in $\mathcal{C}$ (i.e. are a countable intersection of open dense sets). Any path γ joining two generic metrics g_0 and g_1 in $\mathcal{C}$ can be perturbed to be transverse to this projection. Then $\pi^{-1}(\gamma)$ is an oriented manifold of dimension $\dim M_L + 1$ in $B_L \times [0, 1]$, and, provided that none of the moduli spaces $M_L(\gamma(t))$ contain reducible solutions, $\pi^{-1}(\gamma)$ is an oriented cobordism between $M_L(g_0)$ and $M_L(g_1)$. So

$$SW_X(L, g_0) = \langle \beta^n, [M_L(g_0)] \rangle = \langle \beta^n, [M_L(g_1)] \rangle = SW_X(L, g_1)$$

Thus the problem lies with reducible solutions.

The curvature F_A of a connection A on the complex line bundle L is a closed 2-form (by the Bianchi identity) and represents the cohomology class

$$[F_A] = 2\pi c_1(L) \in H^2(X; \mathbf{R}).$$

If A is anti-self-dual with respect to a metric g on X then $d^* F_A = - * d * F_A = *dF_A = 0$. Thus if A is g-anti-self-dual, F_A is g-harmonic. Identify $H^2(X; \mathbf{R})$ with the g-harmonic 2-forms, and let

$$H^2(X; \mathbf{R}) = \mathcal{H}_g^+ \oplus \mathcal{H}_g^-$$

be its decomposition into the ± 1 eigenspaces of the $*$-operator of g. We see that A is g-anti-self-dual if and only if $F_A = 2\pi c_1(L) \in \mathcal{H}_g^-$. The codimension of the vector subspace $\mathcal{H}_g^-$ of $H^2(X;\mathbf{R})$ is b^+; so it is reasonable that if $b^+ \geq 1$, the lattice point $c_1(L)$ will not lie on $\mathcal{H}_g^-$ for a generic metric g, and L will admit no g-anti-self-dual connections. If $b^+ \geq 2$ the same will be true for paths of metrics as in our argument above. Rigorous proofs of these facts can be given by using Sard-Smale theory (cf. [DK]).

If $b^+ \geq 3$, we see that generic paths of generic metrics will admit no reducible solutions of the Seiberg-Witten equations; so SW_X will be a diffeomorphism invariant. For the proof of the Immersed Thom Conjecture, however, we shall be interested in manifolds with $b^+ = 1$. Thus we need to keep track of the metric in our notation: $SW_X(L, g)$. As we have seen above, the line bundle L will admit a g-anti-self-dual connection provided $c_1(L) \in \mathcal{H}_g^-$. We are assuming $b^+ = 1$; so for any metric g, $\mathcal{H}_g^+ \equiv \mathbf{R}$, and there is a unique g-self-dual 2-form, ω_g, up to scale. Since $\mathcal{H}_g^+$ and $\mathcal{H}_g^-$ are L^2-orthogonal, L will admit a g-anti-self-dual connection if and only if $c_1(L) \cdot \omega_g = 0$. The self-dual harmonic 2-form

$$\omega_g \in \mathbf{P}(\{\alpha \in H^2(X;\mathbf{R}) | \alpha \cdot \alpha > 0\})$$

is called the *period point* of the metric g. Note that $\{\alpha \in H^2(X;\mathbf{R}) | \alpha \cdot \alpha > 0\}$ is the positive cone of the intersection form of X. Reducible solutions to the Seiberg-Witten equations can appear only for those metrics g whose period points ω_g lie in the hyperplane $c_1(L)^\perp$. This hyperplane separates $\mathbf{P}(\{\alpha \in H^2(X;\mathbf{R}) | \alpha \cdot \alpha > 0\})$ into two chambers given by the inequalities $c_1(L) \cdot \omega_g > 0$ and $c_1(L) \cdot \omega_g < 0$. Any two generic metrics g_0 and g_1 whose period points lie in the same chamber can be connected by a path of metrics in that chamber, and our argument above shows that $SW_X(L, g_0) = SW_X(L, g_1)$. Thus, for any characteristic line bundle L, the invariants $SW_X(L, g)$ take on at most two possible values as functions of g.

To see what happens as we change metrics via a path γ whose period points cross the hyperplane $c_1(L)^\perp$ transversely at a single point, we can invoke the Kuranishi model of the parametrized moduli space $\pi^{-1}(\gamma)$. (See [DK], [D2], and Lecture 3 for more information about the Kuranishi model.) In case the formal dimension $\dim M_L = 0$, it models the one-parameter family of moduli spaces $M_L(g_t)$ near a reducible solution $(A, 0) \in M_L(g_0)$ as the zero set of the map $z \to |z|^2 + t$ from $\mathbf{C} \to \mathbf{R}$. Thus we see that

$$SW_X(L, g_{-1}) = SW_X(L, g_{+1}) \pm 1$$

depending on the direction that the path crosses the hyperplane. Virtually the same argument extends to the case where $\dim M_L > 0$. Thus we have a simple 'wall-crossing formula'.

We can now proceed with the proof of the theorem. Let us suppose that we have a smooth regular immersion $S^2 \to \mathbf{CP}^2$ which has p positive and n negative double points. Assume also that this immersion represents the class dH and that

$$p = \frac{(d-1)(d-2)}{2} - 1 = \frac{d^2 - 3d}{2} \tag{2}$$

and look for a contradiction. (In case there are fewer positive double points we can always increase the number by connect sums with immersed 2-spheres representing $0 \in H_2$ and with a pair of 'cancelling' double points.)

We next turn our immersion in $\mathbf{CP}^2$ into an embedding in $\mathbf{CP}^2 \# N\overline{\mathbf{CP}}^2$ by blowing up. Here $\overline{\mathbf{CP}}^2$ denotes $\mathbf{CP}^2$ with its opposite orientation. Topologically, the process of blowing up forms the connected sum of a manifold with $\overline{\mathbf{CP}}^2$. A projective line $\overline{\mathbf{CP}}^1 = E \subset \overline{\mathbf{CP}}^2$ is called an *exceptional curve*. Its self-intersection is $E \cdot E = -1$. Blowing up at a double point of our immersion will remove the double point in the connected sum with $\overline{\mathbf{CP}}^2$. This process adds one copy of $\pm E$ to each sheet of the immersion at this point. If the double point is positive both copies are $-E$, and if the double point is negative then one copy is E and the other is $-E$. Thus if $\Sigma \subset X_0$ is represented by an immersed 2-sphere with q double points, then by blowing up at a double point we obtain an immersed sphere with $q - 1$ double points in $X_0 \# \overline{\mathbf{CP}}^2$ representing $\Sigma - 2E \in H_2(X_0 \# \overline{\mathbf{CP}}^2; \mathbf{Z})$ if the double point is positive, and representing Σ if the double point is negative.

Let X be the rational surface obtained by blowing up $\mathbf{CP}^2$ a number $N \geq p+q$ times so that

(a) The homology class

$$\Sigma = dH - 2\sum_{i=1}^{p} E_i - \sum_{j=1}^{q} E_{j+p}$$

is represented by an embedded 2-sphere in $X = \mathbf{CP}^2 \# N\overline{\mathbf{CP}}^2$ (which is diffeomorphic to a rational surface) where $N = p + n + q$, and

(b) q is chosen so that the self intersection $\Sigma \cdot \Sigma = d^2 - 4p - q < 0$.

Recall our assumption (2) that $p = (d^2 - 3d)/2$; so $\Sigma \cdot \Sigma = 6d - d^2 - q$. Now let

$$K = 3H - \sum_{i=1}^{N} E_i$$

which is the negative of the canonical class of the rational surface X. Let us also denote by K the complex line bundle over X whose chern class is K. It is clear that K is characteristic, in fact, $K = c_1(X)$; so it follows that $\dim M_K = 0$. This can also be seen easily by plugging into the dimension formula (1). Another characteristic line bundle is given by $K - 2\Sigma$ (still using additive notation by identifying a line bundle with its chern class), and the dimension formula shows that

$$\dim M_{K-2\Sigma} = \dim M_K + \Sigma \cdot \Sigma - \Sigma \cdot K$$

However by (2)

$$\Sigma \cdot K = 3d - 2p - q = 6d - d^2 - q = \Sigma \cdot \Sigma$$

so also $\dim M_{K-2\Sigma} = 0$. A key point here is that if we were to change our assumption (2) by increasing the right hand side (so that there should not be a contradiction), then the resultant formal dimension of $M_{K-2\Sigma}$ would be negative and the rest of our discussion would not apply.

Since $b^+(X) = 1$, each of the bundles determines a hyperplane $K^\perp$ and $(K-2\Sigma)^\perp$ and corresponding chambers of the projectivization of the positive cone of $H^2(X; \mathbf{R})$. We next wish to determine the Seiberg-Witten invariants $SW_X(K, g)$ and $SW_X(K - 2\Sigma, g)$. Since the Fubini-Study metric on $\mathbf{CP}^2$ (and on $\overline{\mathbf{CP}}^2$) has positive scalar curvature, we can glue these together on the connected sum, as explained in [GL], to obtain a metric of positive scalar curvature on X. If we take a sequence of metrics $\{g_t\}$ shrinking the size of the necks in the connected sum to zero, we obtain the wedge of the Fubini-Study metrics on $\mathbf{CP}^2 \amalg \overline{\mathbf{CP}}^2 \amalg \cdots \amalg \overline{\mathbf{CP}}^2$. The limit of the period points ω_{g_t} is a harmonic self-dual 2-form on $\mathbf{CP}^2 \amalg \overline{\mathbf{CP}}^2 \amalg \cdots \amalg \overline{\mathbf{CP}}^2$. The self-intersection of a self-dual form is nonnegative; so the restriction of the limit of the period points must vanish on each $\overline{\mathbf{CP}}^2$, and thus up to scale it is H. Write g_+ for g_t, t large; so we see that ω_{g_+} is approximately equal to H. Since g_+ has positive scalar curvature, $SW(K, g_+) = 0$ and $SW(K - 2\Sigma, g_+) = 0$. But $K \cdot \omega_{g_+} \sim K \cdot H = 3 > 0$, and $(K - 2\Sigma) \cdot \omega_{g_+} \sim (K - 2\Sigma) \cdot H = 3 - 2d < 0$, provided $d \geq 2$. Since our theorem is clear for $d = 1$, we assume now that $d \geq 2$. The wall-crossing formula thus implies:

$$SW_X(K, g) = \begin{cases} 0, & \text{if } K \cdot \omega_g > 0 \\ \pm 1, & \text{if } K \cdot \omega_g < 0 \end{cases} \tag{3}$$

$$SW_X(K - 2\Sigma, g) = \begin{cases} \pm 1, & \text{if } K \cdot \omega_g > 0 \\ 0, & \text{if } K \cdot \omega_g < 0 \end{cases} \tag{4}$$

To complete our argument we consider a tubular neighborhood $U \subset X$ of the smoothly embedded 2-sphere representing the homology class Σ. The boundary of U is the lens space $L(p, -1)$. We write $X = X_0 \cup L(p, -1) \times (-\epsilon, \epsilon) \cup U$. The neighborhood U admits metrics of positive scalar curvature; so we can obtain a family of generic metrics on X: $g_r = g_0 \cup g_{L,r} \cup g_U$ where $g_{L,r}$ and g_U have positive scalar curvature, and $g_{L,r}$ makes the neck isometric to $L(p, -1) \times (-r, r)$. Let $X_0^+ = X_0 \cup L(p, 1) \times [0, \infty)$ and $U^+ = U \cup L(p, -1) \times [0, \infty)$. The Mayer-Vietoris sequence gives an isomorphism $j^* : H^2(X; \mathbf{R}) \cong H^2(X_0^+; \mathbf{R}) \oplus H^2(U^+; \mathbf{R})$. Let $j^*(K) = K_0 + K_U$. An argument similar to the one given above shows that for the $b^+ = 1$ (metrically) cylindrical end manifold X_0^+, a metric admits an anti-self-dual connection on the line bundle with $c_1 = K_0$ if and only if its period point is orthogonal to K_0, and for a generic metric this condition does not hold. Thus we may choose g_0 so that

$$C = \lim_{r \to \infty} \omega_{g_r}|_{X_0^+} \cdot K_0 \neq 0.$$

Since the necks have positive scalar curvature, there is a gluing theory for obtaining solutions to the Seiberg-Witten equations on X from solutions on X_0^+ and U^+, and this theory parallels the gluing theory for solutions of the anti-self-duality equations on a connected sum. (See [W] for example.)

Since the neighborhood U^+ has positive scalar curvature, the only solution of the Seiberg-Witten equations for K is the reducible solution $(A, 0)$ where A is an anti-self-dual connection on $K|_{U^+}$. Similarly, the only solution on U^+ for $K - 2\Sigma$ is the reducible solution $(A', 0)$. Note that the line bundles K and $K - 2\Sigma$ agree on X_0^+ and so we may identify the moduli spaces $M_{X_0^+}(K) = M_{X_0^+}(K - 2\Sigma) = M_0$. Since

$$0 = \dim M_X(K) = \dim M_{X_0^+}(K) + \dim M_{U^+}(K) + 1$$

and

$$\dim M_X(K - 2\Sigma) = 0,$$

the formal dimensions of the moduli spaces $M_{U^+}(K)$ and $M_{U^+}(K - 2\Sigma)$ are equal. In fact an index calculation using the Atiyah-Patodi-Singer formula shows that both dimensions are equal to -1. It follows from the gluing theory that for the metric g_r, with r large

$$M_{X,g_r}(K) \cong M_0 \times \{(A, 0)\} \cong M_0 \times \{(A', 0)\} \cong M_{X,g_r}(K - 2\Sigma).$$

Hence, counting points in these 0-dimensional moduli spaces we obtain for large r,

$$SW_X(K, g_r) \equiv SW_X(K - 2\Sigma, g_r) \pmod 2. \tag{5}$$

However, since $\Sigma \cdot \Sigma < 0$, the intersection form of U is negative definite, so $\lim_{r \to \infty} \omega_{g_r}|_{U^+} = 0$. This means that

$$\lim_{r \to \infty} \omega_{g_r} \cdot (K - 2\Sigma) = \lim_{r \to \infty} \omega_{g_r} \cdot K = C.$$

Hence for large r, $\omega_{g_r} \cdot (K - 2\Sigma)$ and $\omega_{g_r} \cdot K$ both have the same sign as C. This contradicts (5) and (3) and (4) and completes the proof of the theorem.

As a matter of fact, the Immersed Thom Conjecture can also be proved by following Kronheimer and Mrowka's proof of the classical Thom Conjecture. First remove all the positive double points of the immersion by adding handles to increase the genus by exactly p. Then blow up n times to remove the negative double points and follow the proof of [KM2]. However, notice that the above proof also gives new information about representing homology classes in the rational surface $\mathbf{CP}^2 \# q\overline{\mathbf{CP}}^2$ by immersed spheres.

Corollary. *Let* $\alpha = dH + \sum_1^q a_i E_i \in H_2(\mathbf{CP}^2 \# q\overline{\mathbf{CP}}^2; \mathbf{Z})$. *Then if* $d \geq 2$ *and*

$$d^2 - 3d \geq \sum_1^q (a_i^2 - a_i) + 2p$$

the class α cannot be represented by an immersed 2-sphere with p positive double points.

The question of representability by smoothly embedded 2-spheres ($p = 0$) has been well-studied in the literature in the case where $\alpha \cdot \alpha \geq 0$ where one may apply Donaldson's theorems about the realization of intersection forms which are discussed in the next lecture. See [L] for a survey of results. The chief interest of the corollary is where $\alpha \cdot \alpha < 0$.

Lecture 3: Intersection Forms of Smooth 4-Manifolds

The most basic invariant of a (compact) oriented simply connected smooth 4-manifold is its intersection form. As was mentioned in Lecture 1, it can be defined homologically. If $a, b \in H_2(X; \mathbf{Z})$ then the intersection $a \cdot b$ is obtained by counting signed transverse intersections of oriented surfaces in the given homology classes, or equivalently, by evaluating the cup product of the Poincaré duals on the fundamental class of X. The intersection form Q_X is a unimodular symmetric bilinear form

$$H_2(X; \mathbf{Z}) \otimes H_2(X; \mathbf{Z}) \to \mathbf{Z}$$

(a $\mathbf{Z}$-inner product space). According to a theorem of J.H.C. Whitehead [Wh], Q_X determines the oriented homotopy type of X. One of the most important questions of 4-manifold theory is to determine which $\mathbf{Z}$-inner product spaces are realized as intersection forms Q_X for simply connected X. The first major advances in this direction arose through the work of Simon Donaldson. It is the purpose of this lecture to discuss Donaldson's theorems from the point of view of Seiberg-Witten theory and also new results in this area due to Mikio Furuta. Throughout, we shall take the Seiberg-Witten point of view. We refer the reader to [DK] for an exposition of the results in this area using Donaldson theory.

Integral inner product spaces are distinguished according to parity and definiteness. A $\mathbf{Z}$-inner product space A is *even* if $a \cdot a$ is even for each $a \in A$. Otherwise, A is called *odd*. If $a \cdot a < 0$ ($a \cdot a > 0$) for each $a \in A$, then A is called *negative (positive) definite*. Otherwise A is *indefinite*. Indefinite $\mathbf{Z}$-inner product spaces are completely classified by their rank, signature and parity: If A is odd, then it is a direct sum of the forms

$$A \cong p\,(1) \oplus q\,(-1)$$

and if A is even then

$$A \cong r\,E_8 \oplus n\,H$$

where E_8 is the rank 8 negative definite form

$$E_8 = \begin{pmatrix} -2 & 1 & 0 & 0 & 0 & 0 & 0 & 0 \\ 1 & -2 & 1 & 0 & 0 & 0 & 0 & 0 \\ 0 & 1 & -2 & 1 & 0 & 0 & 0 & 0 \\ 0 & 0 & 1 & -2 & 1 & 0 & 0 & 0 \\ 0 & 0 & 0 & 1 & -2 & 1 & 0 & 1 \\ 0 & 0 & 0 & 0 & 1 & -2 & 1 & 0 \\ 0 & 0 & 0 & 0 & 0 & 1 & -2 & 0 \\ 0 & 0 & 0 & 0 & 1 & 0 & 0 & -2 \end{pmatrix}$$

and H is the indefinite (signature $= 0$) form

$$H = \begin{pmatrix} 0 & 1 \\ 1 & 0 \end{pmatrix}.$$

The notation 'pA' refers to the direct sum of p copies of A.

Definite **Z**-inner product spaces are not so well understood and are classified only for small rank. There are many examples. Although E_8 is the only negative definite **Z**-inner product space of rank 8, there are, for example, thousands of rank 40. Let us first think about the realizability of (say negative) definite forms. For example, $Q_{\overline{\mathbf{CP}}^2} = (-1)$. Since $Q_{X\#Y} \cong Q_X \oplus Q_Y$, we can realize $Q_{q\overline{\mathbf{CP}}^2} \cong q(-1)$. Let us call $q(-1)$ *standard*. These are the only ones that can be obviously realized as negative definite intersection forms of simply connected 4-manifolds. The case $q = 0$ gives Q_{S^4}, the only standard definite form which is even. Donaldson's first application of his remarkable work to the theory of 4-manifolds was to prove that a definite intersection form of a simply connected smooth 4-manifold must be standard. Thus topologists studying smooth 4-manifolds needn't worry that the algebraic study of definite **Z**-inner product spaces proves so difficult.

Theorem (DONALDSON [D1]). *If X is a simply connected smooth 4-manifold whose intersection form is negative definite, then $Q_X \cong b^-(-1)$.*

It should be pointed out that it was this theorem, together with M. Freedman's work on topological 4-manifolds, that proved the existence of the celebrated 'fake $\mathbf{R}^4$'s'. We shall give a proof of Donaldson's theorem using Seiberg-Witten theory. This proof is actually similar in spirit to the proof given in [FS1]. First consider the case where Q_X is even. We need to show that $H_2(X; \mathbf{Z}) = 0$. Note that a simply connected 4-manifold whose intersection form is even admits a spin structure. This is because for such a manifold the Wu class v_2 is the Stiefel-Whitney class w_2 [MS], and any $\alpha \in H_2(X; \mathbf{Z}_2)$ is the mod 2 reduction of an integral class $a \in H_2(X; \mathbf{Z})$. Thus $\langle w_2, a \rangle \equiv a \cdot a \equiv 0 \pmod{2}$, and so $w_2 = 0$. This means that the trivial complex line bundle $\mathbf{C}$ is characteristic. The corresponding Seiberg-Witten moduli space has formal dimension

$$\dim M_{\mathbf{C}} = -\frac{\text{sign}\,(X)}{4} - (1 + b^+) = 2k - 1$$

where k is the index $\hat{A}(X)$ of the Dirac operator on X ($b^+ = 0$). The moduli space $M_{\mathbb{C}}$ is nonempty since it contains the trivial (and reducible) solution $(\Theta, 0)$. We wish to study $M_{\mathbb{C}}$ near this trivial solution.

Note that connections on $\mathbb{C}$ can be identified with 1-forms on X; so we consider the operator $P : \Omega^1 \times \Gamma(V^+) \to \Omega^2_+ \times \Gamma(V^-)$ given by $P(\omega, \psi) = (F^+_\omega - q(\psi), D_\omega \psi)$ which vanishes precisely on the solutions to the Seiberg-Witten equations. Its linearization L is a Fredholm map given by the matrix

$$L = \begin{pmatrix} d^+ & 0 \\ 0 & D \end{pmatrix}$$

which we view as having domain $\ker d^* \oplus \Gamma(V^+)$. Thus $P = L + \Phi$ where Φ is the quadratic map $\Phi(\omega, \psi) = (-q(\psi), \omega.\psi)$ (the '.' in the second coordinate is Clifford multiplication). After splitting off $\ker L$ from the domain and coker L from the range, L becomes an isomorphism. Let p_1 be the projection of $\Omega^2_+ \times \Gamma(V^-)$ orthogonal to coker L, and let p_0 be the complementary projection onto coker L. The map $f : \ker d^* \oplus \Gamma(V^+) \to \ker d^* \oplus \Gamma(V^+)$ defined by $f(u) = u + L^{-1} p_1 \Phi(u)$ defines a change of coordinates near the origin. If the equation $P(u) = 0$ is projected using p_0 and p_1 and then rewritten in terms of coordinates $v = f(u)$, one obtains a finite dimensional local model given as the zero set of a map $\varphi : \ker L \to$ coker L called the Kuranishi map. In fact $\varphi(v) = p_0 P(u)$. Now $\ker L = H^1(X; \mathbf{R}) \oplus \ker D$ and coker $L = H^2_+(X; \mathbf{R}) \oplus$ coker D. In our case, $H^1(X; \mathbf{R}) = 0 = H^2_+(X; \mathbf{R})$, and the index of D is k; so the Kuranishi map is

$$\varphi : \mathbf{C}^{k+r} \to \mathbf{C}^r.$$

Our solution in hand, $(\Theta, 0)$, is reducible, and so has a stabilizer group isomorphic to a copy of S^1 in the gauge group Aut $(\mathbb{C})$. Let $\tilde{B}_{\mathbb{C}}$ denote $\Omega^1 \times \Gamma(V^+)/$ Aut $^0(\mathbb{C})$; so S^1 acts on $\tilde{B}_{\mathbb{C}}$. Its quotient map is the basepoint map, and $(\Theta, 0)$ is a fixed point. The group S^1 acts with the standard weight one representation on $\ker D$ and coker D, and the zero set of φ actually models a neighborhood of $(\Theta, 0) \in \tilde{M}_{\mathbb{C}}$ in a slice to the S^1-action. To get a model of a neighborhood of $(\Theta, 0)$ in $M_{\mathbb{C}}$, we thus need to divide by the stabilizer group S^1. Consider the restriction of φ to a small sphere in $\mathbf{C}^{k+r}$, and view φ as a section of the trivial $\mathbf{C}^r$ bundle over $S^{2(k+r)-1}$. Taking the quotient by S^1 gives a section φ' of the bundle $\bigoplus_r \gamma \to \mathbf{CP}^{k+r-1}$, where γ denotes the canonical line bundle. Since $M_{\mathbb{C}} \setminus \{(\Theta, 0)\}$ is generically a manifold, we may suppose that our choice of φ was generic; so that φ' vanishes transversely. Any two such sections are homotopic, and there is a cobordism of the corresponding zero sets. The section given by projection on the last r homogeneous coordinates gives $\mathbf{CP}^{k-1}$ as zero set.

Let $M'_{\mathbb{C}}$ denote the complement in $M_{\mathbb{C}}$ of the open cone neighborhood of $(\Theta, 0)$. This is a compact $(2k-1)$-manifold with boundary $\mathbf{CP}^{k-1}$. Since a characteristic line bundle has at most one reducible solution to the Seiberg-Witten equations, the restriction of the basepoint map is a fibration

$$\tilde{M}'_{\mathbf{C}} \to M'_{\mathbf{C}}$$

and this restricts over the boundary as the canonical S^1-fibration $S^{2k-1} \to \mathbf{CP}^{k-1}$. The standard generator h of $H^2(\mathbf{CP}^{k-1}; \mathbf{Z})$ is the chern class of the fibration over the boundary, i.e. the restriction of the class β of Lecture 2 to $\partial M'_{\mathbf{C}}$ is just h. So we get the contradiction

$$1 = \langle h^{k-1}, [\mathbf{CP}^{k-1}] \rangle = \langle \beta^{k-1}, \partial [M'_{\mathbf{C}}] \rangle = 0$$

in case $k \geq 1$. This means that $k = -(\,\mathrm{sign}\,(X)/8) = 0$; so $H_2(X; \mathbf{Z}) = 0$, as we claimed.

This same proof works in general; i.e. if we assume that Q_X is negative definite and nonstandard, then provided we are able to find a characteristic line bundle L on X whose Seiberg-Witten moduli space has formal dimension

$$\dim M_L = \frac{\lambda \cdot \lambda - \mathrm{sign}\,(X)}{4} - 1 > 0$$

where $\lambda = c_1(L)$, the same proof gives a contradiction. That this can be done follows from a number-theoretic result of Noam Elkies [E]:

Proposition (ELKIES). *Let A be a negative definite $\mathbf{Z}$-inner product space. Then the shortest characteristic vectors (i.e. w in A such that $v \cdot w = v \cdot v$ (mod 2) for all v in A) have norm at most $-\mathrm{sign}\,(A)$, with equality if and only if A is standard.*

Let us now turn to the case of indefinite forms. Some of these can be realized easily. For example, we can realize all odd indefinite $\mathbf{Z}$-inner product spaces:

$$p(1) \oplus q(-1) \cong Q_{p\mathbf{CP}^2 \# q\overline{\mathbf{CP}}^2}.$$

As for examples of even forms, we have $Q_{S^2 \times S^2} = H$. An important example in 4-manifold topology is the $K3$-surface. We may think of it as the nonsingular degree 4 hypersurface in $\mathbf{CP}^3$,

$$K3 = \{(x, y, z, w) \in \mathbf{CP}^3 | x^4 + y^4 + z^4 + w^4 = 0\}.$$

Its intersection form is $Q_{K3} \cong 2E_8 \oplus 3H$. Thus by taking connected sums of copies of $K3$ and $S^2 \times S^2$ we may realize any even indefinite form of the type $2mE_8 \oplus (3m + n)H$. Recall that any even indefinite $\mathbf{Z}$-inner product space is isomorphic to $rE_8 \oplus nH$. Rohlin's theorem [R] asserts that if $Q_X \cong rE_8 \oplus nH$ (recall we are assuming X is simply connected here) then r is even. This can be seen by noting that X admits a spin structure whose Dirac operator has index equal to $\hat{A}(X) = -\mathrm{sign}\,(X)/8$. The Dirac operator

$$D : \Gamma(V^+) \to \Gamma(V^-)$$

is an operator on bundles with structure group $SU(2) = Sp(1)$; therefore they have quaternionic structures. The Dirac operator preserves these structures; so its index, $\dim_{\mathbf{C}}(\ker D) - \dim_{\mathbf{C}}(\operatorname{coker} D)$ must be even. Thus $\operatorname{sign}(X)$ is divisible by 16, and we may write $Q_X \cong 2mE_8 \oplus nH$. One is interested in the ratio of m to n or equivalently to the ratio of the rank b_2 to the signature. One of the most famous conjectures in 4-manifold theory is:

The 11/8 Conjecture. *If X is a simply connected smooth 4-manifold with Q_X even then*

$$\frac{b_2}{|\operatorname{sign}(X)|} \geq \frac{11}{8}.$$

This conjecture is still unsettled, however we shall discuss a bit later some positive progress reported by Furuta. If we write $Q_X \cong 2mE_8 \oplus nH$, then the 11/8 conjecture is equivalent to the assertion that $n \geq 3|m|$. Equality is realized by the intersection form of the connected sum of any number of copies of $K3$. The first positive progress in the direction of the 11/8 Conjecture was due to Simon Donaldson [D3], where he proved that the conjecture is true for $m = 1$:

Theorem (DONALDSON). *If X is a simply connected smooth 4-manifold whose intersection form is $Q_X \cong 2mE_8 \oplus nH$ with $n \leq 2$, then $m = 0$.*

Donaldson's ideas that lead to the proof of this theorem are indeed beautiful ([D3] is my favorite paper in gauge theory), but the proof is very difficult. Here we outline a proof due to Peter Kronheimer using Seiberg-Witten theory. We shall actually show that there is no X with $Q_X \cong 2E_8 \oplus 2H$. The general case, showing that there is no X with $Q_X \cong 2mE_8 \oplus 2H$, $m > 0$, can be easily reduced to this one, and the case where $n = 1$ follows by taking a connected sum with $S^2 \times S^2$.

So assume that $Q_X \cong 2E_8 \oplus 2H$. Again, the trivial line bundle $\mathbf{C}$ is characteristic,

$$\dim M_{\mathbf{C}} = -\frac{\operatorname{sign}(X)}{4} - (1 + b^+) = 1$$

and $M_{\mathbf{C}}$ contains the trivial solution $(\Theta, 0)$. Since $\operatorname{ind} D = 2$, we see that the Kuranishi model of a neighborhood of $(\Theta, 0)$ in $M_{\mathbf{C}}$ is given by the quotient by S^1 of the zero set of the S^1-equivariant map

$$\varphi : \ker D = \mathbf{C}^{2+r} \longrightarrow \operatorname{coker} D \oplus H_+^2(X; \mathbf{R}) = \mathbf{C}^r \oplus \mathbf{R}^2.$$

After taking the quotient by S^1 (which acts trivially on the $\mathbf{R}$ summands), we may think of φ as a section of the bundle

$$\bigoplus_r \gamma \oplus \mathbb{R}^2$$
$$\varphi \uparrow \quad \downarrow$$
$$c\mathbf{CP}^{1+r} \setminus \{c\}$$

where $c\,\mathbf{CP}^{1+r}$ denotes the cone, with cone point c, and $\mathbb{R}^2$ is a trivial real 2-plane bundle. The local model of $M_{\mathbb{C}}$ is the zero set of φ. Again, let $M'_{\mathbb{C}}$ denote the complement in $M_{\mathbb{C}}$ of its intersection with the open cone neighborhood of $(\Theta, 0)$. Then $M'_{\mathbb{C}}$ is a compact 1-dimensional manifold with boundary. One might hope to find a contradiction as in [FS1] by showing that $M'_{\mathbb{C}}$ has an odd number of endpoints. The number of endpoints counted with sign is the number of zeros of the restriction of φ to $\mathbf{CP}^{1+r}$. This is by definition the euler number of the bundle $\bigoplus_r \gamma \oplus \mathbb{R}^2$ over $\mathbf{CP}^{1+r}$; i.e. 0. Thus the number of endpoints is even, and this technique doesn't work.

There is, however, an aspect of $M_{\mathbb{C}}$ which we have not yet used, namely, that since the spinor bundles $V^{\pm}$ are quaternionic line bundles, there is a $\mathbf{Z}_4$-action on them given by the action of the standard quaternion J. (This quaternionic structure was used above in our outline of the proof of Rohlin's theorem. Also note that this means that the complex dimension r of coker D is even, $r = 2s$.) There is then a $\mathbf{Z}_4$-action on $\Omega^1 \times \Gamma(V^+)$ given by $(A, \psi) \to (-A, J\psi)$. Since S^1 is central in Aut $(\mathbb{C})$, it fixes connections A but acts on ψ, and in the quotient (A, ψ) becomes equivalent to $(A, -\psi)$; so the $\mathbf{Z}_4$ action descends to an involution τ on $B_{\mathbb{C}}$. It is not difficult to see that the fixed points of τ are precisely the pairs $(A, 0)$. Thus τ acts freely on $B_{\mathbb{C}}^* \simeq \mathbf{CP}^\infty$. One can compute the mod 2 homology of $\mathbf{CP}^\infty/\tau$ by means of the $\mathbf{Z}_2$-Gysin sequence associated to the real line bundle ξ corresponding to τ. One obtains

$$H_i(\mathbf{CP}^\infty/\tau; \mathbf{Z}_2) = \begin{cases} \mathbf{Z}_2 = \langle e_4{}^k \rangle & i = 4k \\ \mathbf{Z}_2 = \langle e_4{}^k w_1 \rangle & i = 4k+1 \\ \mathbf{Z}_2 = \langle e_4{}^k w_1{}^2 \rangle & i = 4k+2 \\ 0 & i = 4k+3 \end{cases} \tag{6}$$

where $w_1 = w_1(\xi)$ and e_4 is a 4-dimensional class satisfying $\pi^*(e_4) = h^2$ where $\pi : \mathbf{CP}^\infty \to \mathbf{CP}^\infty/\tau$. Now we can divide our local model by τ:

$$\begin{array}{ccc} \bigoplus_{2s} \gamma \oplus \mathbb{R}^2 & & \bigoplus_s (\gamma \oplus \gamma)/\tau \oplus \xi \oplus \xi \\ {\scriptstyle \varphi}\uparrow\downarrow & \xrightarrow{\ /\tau\ } & \downarrow\uparrow{\scriptstyle \bar\varphi} \\ \mathbf{CP}^{1+2s} & & \mathbf{CP}^{1+2s}/\tau \end{array}$$

Since τ acts freely on $M'_{\mathbb{C}}$, the quotient $M'_{\mathbb{C}}/\tau$ is a compact 1-manifold with boundary $\bar\varphi^{-1}(0)$. Its mod 2 number of boundary points is given by the mod 2 euler number of the bundle

$$\bigoplus_s (\gamma \oplus \gamma)/\tau \oplus \xi \oplus \xi \to \mathbf{CP}^{1+2s}/\tau.$$

This is $\langle w_{4s+2}(\bigoplus_s (\gamma \oplus \gamma)/\tau \oplus \xi \oplus \xi), [\mathbf{CP}^{1+2s}/\tau] \rangle = \langle e_4{}^s w_1{}^2, [\mathbf{CP}^{1+2s}/\tau] \rangle \equiv 1$ (mod 2). Since a compact 1-manifold has an even number of endpoints, we have our contradiction.

The point of Kronheimer's proof is that even though $M'_{\mathbb{C}}$ has an even number of endpoints, the ambient quaternionic structure forces these endpoints to occur in pairs — of which there are an odd number. To prove the general case of the theorem, i.e. that there is no smooth simply connected X with $Q_X = 2mE_8 \oplus 2H$ we can follow the same argument getting a τ-equivariant section of the bundle $\bigoplus_{2s} \gamma \oplus \mathbb{R}^2 \to \mathbf{CP}^{2m-1+2s}$. In this case the moduli space $M_{\mathbb{C}}$ has dimension $4m - 3$. It is easy to see from the Gysin sequence used above that $\pi^* : H^{4i}(\mathbf{CP}^\infty/\tau; \mathbf{Z}_2) \to H^{4i}(\mathbf{CP}^\infty; \mathbf{Z}_2)$ is an isomorphism for all i. The contradiction is now obtained from computing in Z_2 homology

$$
\begin{aligned}
0 &= \langle w_{4s+2}(\bigoplus_s(\gamma \oplus \gamma)/\tau \oplus \xi \oplus \xi), \partial((\pi^*)^{-1}h^{2m-2} \cap [M'_{\mathbb{C}}/\tau])\rangle \\
&= \langle w_{4s+2}(\bigoplus_s(\gamma \oplus \gamma)/\tau \oplus \xi \oplus \xi), \partial(e_4^{m-1} \cap [M'_{\mathbb{C}}/\tau])\rangle \\
&= \langle e_4^{m-1+s}w_1^2, [\mathbf{CP}^{2m-1+2s}/\tau]\rangle = 1,
\end{aligned}
$$

as above.

One might hope that these techniques would extend to prove the 11/8 Conjecture for $m > 1$, however Peter Kronheimer has shown that they do not. On the other hand, recent work of Mikio Furuta [F] has shown that they can be extended to prove a 5/4 Conjecture; i.e. that $n \geq 2|m| + 1$ if X is a simply connected smooth 4-manifold with $Q_X = 2mE_8 \oplus nH$. We will present here a very short description of Furuta's work.

Recall the decomposition

$$
P = L \oplus \Phi : V = \ker d^* \oplus \Gamma(V^+) \to W = \Omega_+^2 \times \Gamma(V^-)
$$

and the Kuranishi map $\varphi : \ker L \to \operatorname{coker} L$. Consider the splittings $V = \bigoplus_{\lambda \geq 0} V_\lambda$ and $W = \bigoplus_{\lambda \geq 0} W_\lambda$ where V_λ and W_λ are the λ-eigenspaces of the operators L^*L and LL^*. Generalizing the Kuranishi construction, let L_Λ be the restriction

$$
L_\Lambda : \bigoplus_{\lambda > \Lambda} V_\lambda \to \bigoplus_{\lambda > \Lambda} W_\lambda, \qquad p_\Lambda \text{ the projection} \quad p_\Lambda : W \to \bigoplus_{\lambda > \Lambda} W_\lambda,
$$

and $\bar{p}_\Lambda$ the complementary projection

$$
\bar{p}_\Lambda : W \to \bigoplus_{\lambda \leq \Lambda} W_\lambda.
$$

Now define $f_\Lambda : V \to V$ by $f_\Lambda(u) = u + L_\Lambda^{-1}p_\Lambda\Phi(u)$. In our discussion of the Kuranishi model, we defined a map f which gives the local change of coordinates. In the current notation, $f = f_0$. Furuta defines the extended Kuranishi map

$$
\varphi_\Lambda : \bigoplus_{\lambda \leq \Lambda} V_\lambda \to \bigoplus_{\lambda \leq \Lambda} W_\lambda
$$

by $\varphi_\Lambda(v) = \bar{p}_\Lambda P(u)$ where $v = f_\Lambda(u)$. This map is equivariant under the actions of both S^1 and J. Using the fact that the Seiberg-Witten moduli space $M_\mathbf{C}$ is compact, Furuta shows:

Proposition (FURUTA). *For $\Lambda \gg 0$, the Kuranishi map φ_Λ gives a global model for the moduli space $M_\mathbf{C}$, i.e. the map φ_Λ is proper and $\varphi_\Lambda^{-1}(0) \cong M_\mathbf{C}$.*

Furuta has made use of this global Kuranishi map as follows. If the intersection form $2mE_8 \oplus nH$ is realized by a smooth 4-manifold, then there is an extended Kuranishi map $\varphi_\Lambda : \mathbf{C}^{2(m+r)} \oplus \mathbf{R}^s \to \mathbf{C}^{2s} \oplus \mathbf{R}^{n+s}$ for large enough r and s, and φ_Λ is S^1 and J equivariant, and $\varphi_\Lambda^{-1}(0)$ is compact. Thus, there is an induced equivariant map of spheres which is obtained by taking such a large radius in the domain that φ_Λ has no zeros on the corresponding sphere. In this way, Furuta produces a stable homotopy class of S^1 and J-equivariant maps of spheres, i.e. an element of

$$\mathcal{C}(m,n) = \varinjlim_{r,s}[S(\mathbf{C}^{2(m+r)} \oplus \mathbf{R}^s), S(\mathbf{C}^{2r} \oplus \mathbf{R}^{n+s})]^{\langle S^1, J\rangle}.$$

If $\mathcal{C}(m,n)$ were empty for $3m > n$, the 11/8 Conjecture would follow by contradiction. However, Kronheimer has nontrivial examples of elements of $\mathcal{C}(5,14)$; so this approach cannot be used to prove the 11/8 Conjecture. On the other hand, Furuta has been able to show (using K-theory) that $\mathcal{C}(m,n)$ is empty when $2m \geq n$. This leads to a '5/4-theorem'.

Theorem (FURUTA). *If X is a simply connected smooth 4-manifold with $Q_X = 2mE_8 \oplus nH$, then $n \geq 2|m| + 1$.*

We shall give a proof of the slightly weaker result $n \geq 2|m|$. We learned this proof from notes of a lecture of Dan Freed. It arises from a slight modification of Furuta's ideas. Since we are mainly going to be concerned with the action of J, we rewrite the extended Kuranishi map as

$$\varphi_\Lambda : \mathbf{H}^{|m|+r} \oplus \mathbf{R}^s \to \mathbf{H}^r \oplus \mathbf{R}^{n+s}$$

where $\mathbf{H}$ is the algebra of quaternions. Let E_0 and E_1 be the vector spaces which are the complexifications of the domain and the range. Then there is induced a proper map $\varphi_\Lambda : E_0 \to E_1$, and φ_Λ is G-equivariant, where $G = \langle S^1, J\rangle$ is the subgroup of $SU(2)$ generated by S^1 and J. View E_i as complex vector bundles over a point with projections π_i and zero sections s_i. We get an induced map in equivariant K-theory

$$\varphi_\Lambda^* : K_G(B(E_1), S(E_1)) \to K_G(B(E_0), S(E_0))$$

where $(B(E_i), S(E_i))$ are the (ball,sphere) pairs. For the point, $K_G(*)$ is the representation ring $R(G)$. There are canonical elements, the Bott classes $\lambda_{E_i} \in$

$K_G(B(E_i), S(E_i))$, which give rise to the Thom isomorphism

$$R(G) \rightarrow K_G(B(E_i), S(E_i))$$
$$\rho \rightarrow \pi_i^*(\rho) \cdot \lambda_{E_i}$$

The pullback

$$s_i^*(\lambda_{E_i}) = \sum (-1)^k \Lambda^k(E_i) = \Lambda_{-1}(E_i) \in R(G).$$

Because of the Thom isomorphism, $\varphi_\Lambda^*(\lambda_{E_1}) = \pi_0^*(\rho) \cdot \lambda_{E_0}$ for some $\rho \in R(G)$. Now

$$\rho \cdot \Lambda_{-1}(E_0) = \rho \cdot s_0^* \lambda_{E_0} = s_0^*(\pi_0^*(\rho) \cdot \lambda_{E_0}) = s_0^*(\varphi_\Lambda^* \lambda_{E_1}) = s_1^* \lambda_{E_1} = \Lambda_{-1}(E_1).$$

Consider the trace χ_i of J in the virtual representation $\Lambda_{-1}(E_i)$. We have

$$\text{tr}_J(\Lambda_{-1}(E_1)) = \text{tr}_J(\rho \cdot \Lambda_{-1}(E_0)) = \text{tr}_J(\rho)\ \text{tr}_J(\Lambda_{-1}(E_0))$$

so χ_0 divides χ_1. Furthermore, $\chi_i = \det(\text{Id} - J)$. Recalling that the action of J on $\mathbf{H}$ is by right multiplication by j:

$$J(z + jw) = -\bar{w} + j\bar{z},$$

we see that on $\mathbf{H} \otimes_{\mathbf{R}} \mathbf{C}$ its matrix is

$$\begin{pmatrix} 0 & 0 & 1 & 0 \\ 0 & 0 & 0 & -1 \\ -1 & 0 & 0 & 0 \\ 0 & 1 & 0 & 0 \end{pmatrix}.$$

It follows that $\det((\text{Id} - J)|_{\mathbf{H} \otimes \mathbf{C}} = 4$. The action of J on $\mathbf{R} \otimes \mathbf{C}$ is given by $J(x) = -x$; so $\det((\text{Id} - J)|_{\mathbf{R} \otimes \mathbf{C}} = 1 - (-1) = 2$. Since χ_0 divides χ_1, we have that $4^{|m|+r} 2^s$ divides $4^r 2^{n+s}$, i.e. $n \geq 2|m|$.

References

[D1] S. Donaldson, *An application of gauge theory to the topology of 4-manifolds*, Jour. Diff. Geom. **18** (1983), 269–316.

[D2] S. Donaldson, *Irrationality and the h-cobordism conjecture*, Jour. Diff. Geom. **26** (1987), 141–168.

[D3] S. Donaldson, *Connections, cohomology and the intersection forms of 4-manifolds*, Jour. Diff. Geom. **24** (1986), 275–341.

[D4] S. Donaldson, *Polynomial invariants for smooth 4-manifolds*, Topology **29** (1990), 257–315.

[DK] S. Donaldson and P. Kronheimer, "The Geometry of Four-Manifolds", Oxford Mathematical Monographs, Oxford University Press, Oxford, 1990.

[E]   N. Elkies, *A characterization of the $\mathbf{Z}_n$ lattice*, preprint.

[FS1]   R. Fintushel and R. Stern, *SO(3) connections and the topology of 4-manifolds*, Jour. Diff. Geom. **20** (1984), 523–539.

[FS2]   R. Fintushel and R. Stern, *The blowup formula for Donaldson invariants*, Annals of Math. (to appear).

[FS3]   R. Fintushel and R. Stern, *Donaldson invariants of 4-manifolds with simple type*, J. Diff. Geom. (to appear).

[FS4]   R. Fintushel and R. Stern, *Rational blowdowns of smooth 4-manifolds*, preprint.

[FS5]   R. Fintushel and R. Stern, *Immersed spheres in 4-manifolds and the immersed Thom conjecture*, preprint.

[Fr]   M. Freedman, *The topology of four-dimensional manifolds*, Jour. Diff. Geom. **17** (1982), 357–454.

[F]   M. Furuta, *An invariant of spin 4-manifolds and the 11/8-conjecture*, Trieste lecture, March 1995.

[HH]   F. Hirzebruch and H. Hopf, *Felder von Flächenelementen in 4-dimensionalen Mannigfaltigkeiten*, Math. Ann. **136** (1958), 156–172.

[KM1]   P. Kronheimer and T. Mrowka, *Embedded surfaces and the structure of Donaldson's polynomial invariants*, J. Diff. Geom. (to appear).

[KM2]   P. Kronheimer and T. Mrowka, *The genus of embedded surfaces in the projective plane*, Math. Research Letters **1** (1994), 797–808.

[GL]   M. Gromov and B. Lawson, *The classification of simply connected manifolds of positive scalar curvature*, Ann. of Math. **111** (1980), 423–434.

[L]   T. Lawson, *Smooth embeddings of 2-spheres in 4-manifolds*, Expo. Math. **10** (1992), 289–309.

[M]   R. Mandelbaum, Four-dimensional topology: an introduction, Bull. Amer. Math. Soc. **2**, (1980), 1–159.

[MS]   J. Milnor and J. Stasheff, "Characteristic Classes", Annals of Mathematics Studies **76**, Princeton University Press, 1974.

[R]   V.Rohlin, *New results in the theory of four dimensional manifolds*, Dok. Akad. Nauk. USSR **84** (1952), 221–224.

[SW1]   N. Seiberg and E. Witten, *Electric-magnetic duality, monopole condensation, and confinement in $N = 2$ supersymmetric Yang-Mills theory*, Nucl. Phys. B426 (1994), 19–52.

[SW2]   N. Seiberg and E. Witten, *Monopoles, duality and chiral symmetry breaking in $N = 2$ supersymmetric QCD*, Nucl. Phys. B431 (1994), 581–640.

[T]   C. Taubes, *The Seiberg-Witten invariants and symplectic forms*, Mathematical Research Letters **1** (1994), 809–822.

[Wa]   C.T.C. Wall, *On simply connected 4-manifolds*, J. London Math. Soc. **39** (1964), 141–149.

[Wh]   J.H.C. Whitehead, *On simply connected 4-dimensional polyhedra*, Comm. Math. Helvetici **45** (1949), 48–92.

[W]   E. Witten, *Monopoles and four-manifolds*, Math. Research Letters **1** (1994), 769–796.

Progress in Nonlinear Differential Equations
and Their Applications, Vol. 29
© 1997 Birkhäuser Verlag Basel/Switzerland

On the Regularity of Classical Field Theories in Minkowski Space-Time R^{3+1}

S. KLAINERMAN

Department of Mathematics, Princeton University
Princeton, NJ 08544

ABSTRACT. One of the central issues in the theory of nonlinear partial differential equations is that of regularity or break-down of solutions to the important physical examples. This is intimately tied to the basic mathematical question of what we actually mean by solutions and, from a physical point of view, to the understanding of the very limits of validity of the corresponding physical theories.

In my lectures I will consider nonlinear wave equations derivable from a relativistic Lagrangean and attempt to approach this very difficult problem from the local point of view of the optimal regularity for which the initial value problem is well posed. I will try to demonstrate that the question is deep and leads an entire new class of estimates for the linear wave equations, related to the well-known Strichartz inequalities. The results will be applied to the Wave Maps and Yang-Mills equations.

Introduction

After an era dominated to a large extent by the development of general methods, mainly for linear equations, when the relative simplicity and richness of the important special cases were often sacrificed in favor of artificial generality, the subject of partial differential equations is now well into a period of renewed interest in its basic examples arising in Geometry and Physics. Yet despite the fact that each important system of equations lies at the very foundation of an entire geometric or physical theory, they all share certain important common characteristics. The most fundamental among them is the fact that they are all, with few exceptions, derived, or derivable from a Variational Principle. Their Lagrangean is invariant under certain basic groups of transformations intimately tied to the nature of the Geometric or Physical Theory described. By Noether's Theorem these symmetries translate into Laws of Conservation. The methods developed for treating such problems take into account the specific geometric structure and special form of the equations in a decisive way. While this last fact is maybe more visible in regard to purely geometric problems like Harmonic Maps, elliptic Yang-Mills, Yamabe Problem etc, I hope to convince you in these lectures that a similar development is taking place in regard to evolution problems, more specifically in Relativistic Field Equations

such as Wave Maps, Yang-Mills equations in Minkowski space-time and General
Relativity.

In the case of systems evolving in time, the time translation symmetry of the
Lagrangean corresponds to the basic law of Conservation of Energy. An important
physical restriction on the type of systems allowed is that the energy be positive.
In the case of one dimensional systems, i.e. ordinary differential equations, the
positivity of the energy leads immediately to a trivial but important consequence.
Namely there can be no finite time break-down of solutions. Since nothing of
consequence can happen in finite time, one can proceed to address the deeper
issues concerning the long time behavior of solutions. It is well known that this
problem is very complicated; interesting physical systems, like the n-body problem,
can exhibit extreme sensitivity to initial conditions, nowadays called chaos. Such
phenomena can, of course, happen also in higher dimensional systems. However,
for these, the mere positivity of the energy no longer excludes the possibility of
finite time break-down.

Take the well known example of the Burger equation,

$$\partial_t u + u u_x = 0.$$

Here u is a function of two variables t, x; assume it is perfectly smooth, and say
compactly supported, at $t = 0$. The total energy of a solution is given by the
integral,

$$\mathcal{E}(t) = \int |u(t, x)|^2 dx.$$

One easily checks that $\mathcal{E}(t) = \mathcal{E}(0)$. In fact even more is true, we can check that,
for every positive integer k,

$$\int |u(t, x)|^{2k} dx = \int |u(0, x)|^{2k} dx.$$

Despite the presence of all these positive conserved quantities all solutions, with
smooth compactly supported nonzero initial data at $t = 0$, break-down in finite
time. The break-down corresponds, physically, to the formation of a shock wave.

Singularities are also known to form, in some special cases, for solutions to
the Einstein field equations in General Relativity. Moreover one expects this to
happen, in general, in the presence of strong gravitational fields. It is also widely
expected that the general solutions of the incompressible Euler equations in three
space dimensions, modeling the behavior of inviscid fluids, break-down in finite
time. Some speculate that the break-down may have something to do with the
onset of turbulence for fluids with very high Reynolds numbers[1].

---

1)   These fluids are in fact described by the Navier-Stokes equations. In this case the
     general consensus is that all smooth solutions remain smooth for all times. The
     problem is still open.

The problem of break-down of solutions to interesting, non-linear, physical systems is clearly one of the central problem in P.D.E's today. It is intimately tied to the basic mathematical question of understanding what we actually mean by solutions and, from a physical point of view, to the issue of understanding the very limits of validity of the corresponding physical theories. Thus in the case of the Burger equation, for example, the problem of singularities can be tackled by extending our concept of solutions to accommodate "shock waves" , i.e solutions discontinuous across curves in the t, x space. One can define a functional space of generalized solutions in which the initial value problem can be uniquely solved. Though the situation for more realistic physical systems is far less clear and far from being satisfactory solved, the generally held opinion here is that shock wave type singularities can be accommodated without breaking the boundaries of the physical theory at hand.

The situation of singularities in General Relativity is radically different. The type of singularities expected here is such that no continuation of the solution is possible without altering the physical theory itself. The prevaling opinion, in this respect, is that only a quantum field theory of Gravity could achieve this.

1. Relativistic Field Theories

In these lectures I will restrict myself to a discussion of the issue of regularity of solutions to Relativistic Field Theories, in other words theories which arise, by the Variational Principle, from a Relativistic Lagrangean.

The underlying geometry of a Relativistic Field Theory, which itself becomes a dynamical variable in General Relativity, is that given by a space-time. This consists of a pair $(\mathcal{M}, \mathbf{g})$ with $\mathcal{M}$ a smooth $(n+1)$-dimensional manifold and $\mathbf{g}$ an Einsteinian metric, that is a nondegenerate quadratic form of signature $(-1, 1, \ldots, 1)$, defined on the tangent space $T_p(\mathcal{M})$ and varying smoothly with $p \in \mathcal{M}$. Thus, locally, a space-time looks Minkowskian. The Minkowski space-time itself, which will be denoted by $\mathbf{M}^{n+1}$, consists of a copy of $\mathbf{R}^{n+1}$ together with a metric $\mathbf{m}$ and a distinguished class of global coordinates x^α, $\alpha = 0, 1, \ldots, n$, relative to which the line element of the metric takes the form,

$$ds^2 = -(dx^0)^2 + (dx^1)^2 + \ldots + (dx^n)^2.$$

We split x^α into the time variable $x^0 = t$ and the space variables $x = (x^1, \ldots, x^n)$.

A relativistic field theory on $\mathcal{M}$ is specified by a Lagrangean L which depends on the metric $\mathbf{g}$ and also on a collection of fields $\mathbf{H} = (\mathbf{H}_1 \ldots, \mathbf{H}_m)$ corresponding to some matter- fields present in the space-time. One defines the action integral $\mathcal{S}[\mathbf{H}, \mathbf{g}] = \mathcal{S}[\mathbf{H}, \mathbf{g}; \mathcal{U}] = \int_{\mathcal{U}} L d\vartheta_{\mathbf{g}}$ where $\mathcal{U}$ is a given relatively compact domain in $\mathcal{M}$ and $d\vartheta_{\mathbf{g}}$ is the volume element of $\mathbf{g}$. A compact variation of the fields $\mathbf{H}$ and metric $\mathbf{g}$ is, by definition, a smooth 1-parameter family $(\mathbf{H}(s), \mathbf{g}(s))$ with $s \in (-\epsilon, \epsilon)$ such that $\mathbf{H}(0) = \mathbf{H}$, $\mathbf{g}(0) = \mathbf{g}$ everywhere in $\mathcal{M}$ and $\mathbf{H}(s) = \mathbf{H}$, $\mathbf{g}(s) = \mathbf{g}$ at all points p in $\mathcal{M} \setminus \mathcal{U}$. The field equations of a particular theory are obtained by requiring that the acceptable physical solutions of the theory are those stationary relative to arbitrary compact variations of the corresponding action integral. Here are some simple examples:

Example 1. Scalar Field Equations
These are obtained by considering a real scalar ϕ defined on $(\mathcal{M}, \mathbf{g})$ with Lagrangean density $L[\phi] = -\frac{1}{2} \mathbf{g}^{\mu\nu} \partial_\mu \partial_\nu \phi - V(\phi)$ with V a given real function of ϕ. The corresponding Euler-Lagrange equations, obtained by varying ϕ alone[2], are given by

$$\Box_{\mathbf{g}} \phi - V'(\phi) = 0 \qquad\qquad \text{N.W.E}$$

where $\Box_{\mathbf{g}}$ is the D'Alembertian operator, $\Box_{\mathbf{g}} = \frac{1}{\sqrt{|-\mathbf{g}|}} \partial_\mu (\mathbf{g}^{\mu\nu} \sqrt{|-\mathbf{g}|}\, \partial_\nu \phi), |-\mathbf{g}| = det(-g_{\mu\nu})$. The equation N.W.E is a non-linear scalar wave equation with potential $V(\phi)$. It has an energy-momentum tensor given by the symmetric 2-tensor,

$$\mathbf{T}_{\alpha\beta} = \frac{1}{2} \left(\phi_{,\alpha} \phi_{,\beta} - \frac{1}{2} \mathbf{g}_{\alpha\beta} (\mathbf{g}^{\mu\nu} \phi_{,\mu} \phi_{,\nu} + 2V(\phi)) \right),$$

where $\phi_\alpha = \partial_\alpha \phi$. It verifies the local conservation laws,

$$\mathbf{D}^\beta \mathbf{T}_{\alpha\beta} = 0. \qquad\qquad (1.1)$$

Moreover, if $V \geq 0$, $\mathbf{T}$ verifies the "positive energy condition" which means that, for any non-vanishing field ϕ, and any timelike, future oriented vectorfields[3] X, Y,

$$\mathbf{T}(X, Y) > 0. \qquad\qquad (1.2)$$

Example 2. Wave Maps
Consider a space-time $(\mathcal{M}, \mathbf{g})$ and a Riemannian manifold $\mathcal{N}$ with metric h. Let ϕ be a mapping from $\mathcal{M}$ to $\mathcal{N}$. We define the Lagrangean density for wave maps to be $L = -\frac{1}{2} Tr_{\mathbf{g}}(\phi^* h)$ where $Tr_{\mathbf{g}}(\phi^* h)$ denotes the trace relative to $\mathbf{g}$ of the pullback $\phi^* h$ of the metric h. In local coordinates x^α, $\alpha = 0, 1, \ldots, n$, on $\mathcal{M}$ and y^a, $a = 1, \ldots, m$ the Lagrangean density takes the form $L = -\frac{1}{2} \mathbf{g}^{\mu\nu} h_{ab}(\phi) \frac{\partial \phi^a}{\partial x^\mu} \frac{\partial \phi^b}{\partial x^\nu} = -\frac{1}{2} \mathbf{g}^{\mu\nu} \langle \phi_\mu, \phi_\nu \rangle_h$, with $\langle \ , \ \rangle_h$ the Riemannian metric on $\mathcal{N}$. The corresponding Euler-Lagrange equations, obtained by variations of $\mathcal{S}$ with respect to ϕ alone, are given by,

$$\Box_{\mathbf{g}} \phi^a + \Gamma^a_{bc}(\phi) \mathbf{g}^{\mu\nu} \partial_\mu \phi^b \partial_\nu \phi^c = 0 \qquad\qquad \text{(W.M.)}$$

where Γ^a_{bc} are the Christoffel symbols of $(\mathcal{N}, h)$.
    The energy-momentum tensor of (W.M.) is given by the symmetric 2-tensor,

$$\mathbf{T}_{\alpha\beta} = \frac{1}{2} \left(\langle \phi_{,\alpha}, \phi_{,\beta} \rangle - \frac{1}{2} \mathbf{g}_{\alpha\beta} (\mathbf{g}^{\mu\nu} \langle \phi_{,\mu} \phi_{,\nu} \rangle) \right).$$

It verifies the local conservation laws 1.1. and the positivity condition 1.2.

---

2)   while the metric $\mathbf{g}$ is kept fixed

3) X is time-like if $\mathbf{g}(X, X) < 0$.

Example 3. Yang-Mills Equations

We first recall the Maxwell equations. Given a $3+1$ dimensional space-time $(\mathcal{M}, \mathbf{g})$ an electromagnetic field $\mathbf{F}$ is a 2-form on $\mathcal{M}$ which is exact, i.e. $\mathbf{F} = d\mathbf{A}$ with $\mathbf{A}$ a 1-form on $\mathcal{M}$. The vector potential $\mathbf{A}$ is unique up to a gauge transformation $\mathbf{A} \mapsto \mathbf{A} + d\chi$ with χ an arbitrary scalar function. The Lagrangean density of $\mathbf{F}$ is defined by $L[\mathbf{F}] = -\frac{1}{2}\mathbf{F}_{\mu\nu}\mathbf{F}^{\mu\nu}$ and the Euler-Lagrange equations are $\mathbf{D}^\nu \mathbf{F}_{\mu\nu} = 0$ with $\mathbf{D}$ denoting the covariant differentiation on $(\mathcal{M}, \mathbf{g})$. Together with the equations $d\mathbf{F} = 0$, which are a consequence of $\mathbf{F} = d\mathbf{A}$, they form the Maxwell equations. We can write the Maxwell equations in a more symmetric form by setting $^*\mathbf{F}_{\mu\nu} = \frac{1}{2}\epsilon_{\mu\nu\alpha\beta}\mathbf{F}^{\alpha\beta}$ the Hodge dual of $\mathbf{F}$ with $\epsilon_{\alpha\beta\gamma\delta}$ the components of the volume form on $(\mathcal{M}, g)$. With this notation the Maxwell equations take the form

$$d\mathbf{F} = 0, \quad d^*\mathbf{F} = 0 \tag{M}$$

The energy-momentum tensor of the Maxwell equations is given by the symmetric 2-tensor

$$\mathbf{T}_{\alpha\beta} = \mathbf{F}^\mu_\alpha \mathbf{F}_{\beta\mu} - \frac{1}{4}g_{\alpha\beta}\,\mathbf{F}_{\mu\nu}\mathbf{F}^{\mu\nu} = \frac{1}{2}\left(\mathbf{F}^\mu_\alpha \mathbf{F}_{\beta\mu} + ^*\mathbf{F}^\mu_\alpha \,^*\mathbf{F}_{\beta\mu}\right).$$

The Yang-Mills equations give a simple generalization of the Maxwell equations where the vector potential $\mathbf{A} = \mathbf{A}_\mu dx^\mu$ becomes a $\mathfrak{g}$-valued 1-form with $\mathfrak{g}$ the Lie algebra of a Lie group $\mathcal{G}^4$. Let $[\,,\,]$ denote the Lie bracket on $\mathcal{G}$ and $\langle\cdot,\cdot\rangle$ its Killing scalar product which one assumes positive definite. For any $\mathcal{G}$-valued tensor $\mathbf{H}$ one defines the gauge covariant derivative $D_\mu \mathbf{H} = \mathbf{D}_\mu \mathbf{H} + [\mathbf{A}_\mu, \mathbf{H}]$ where $\mathbf{D}$ is the covariant derivative on $\mathcal{M}$. The definition is invariant with respect to "gauge transformations". These are transformations of the type

$$\begin{aligned} \mathbf{H}(p) &\mapsto O^{-1}(p)\mathbf{H}(p)O(p) \\ \mathbf{A}_\alpha(p) &\mapsto O^{-1}(p)\mathbf{A}_\alpha(p)O(p) + O^{-1}(p)\partial_\alpha O(p). \end{aligned}$$

One can easily check that for any $\mathcal{G}$-valued scalar $\mathbf{H}$, $D_\mu D_\nu \mathbf{H} - D_\nu D_\mu \mathbf{H} = [\mathbf{F}_{\mu\nu}, \mathbf{H}]$ where the 2-form $\mathbf{F}_{\mu\nu}dx^\mu dx^\nu$, given by the formula

$$\mathbf{F}_{\mu\nu} = \partial_\mu \mathbf{A}_\nu - \partial_\nu \mathbf{A}_\mu + [\mathbf{A}_\mu, \mathbf{A}_\nu], \tag{F}$$

is called the curvature of the connection. It is manifestly invariant under a gauge transformation $\mathbf{F} \to O^{-1}\mathbf{F}O$ and satisfies the Bianchi equation

$$D_\alpha \mathbf{F}_{\beta\gamma} + D_\beta \mathbf{F}_{\gamma\alpha} + D_\gamma \mathbf{F}_{\alpha\beta} = 0.$$

---

4)   For simplicity we shall assume that both $\mathcal{G}$ and $\mathfrak{g}$ are represented by matrices. In fact we can restrict our attention to the classical groups $\mathcal{G} = SO(n, \mathbf{R})$ or $\mathcal{G} = SU(n, \mathbf{C})$ and their corresponding Lie algebras. In this cases the Lie bracket is the standard one $[A, B] = A \cdot B - B \cdot A$ with $\cdot$ the standard matrix multiplication.

The Lagrangean of the Yang-Mills equations is defined by $L[A] = -\frac{1}{4} \langle \mathbf{F}_{\mu\nu}, \mathbf{F}^{\mu\nu} \rangle$. The corresponding Euler-Lagrange equations are

$$D^\nu \mathbf{F}_{\mu\nu} = 0 \qquad\qquad\qquad (\text{Y-M})$$

which together with (F) form the Yang-Mills equations. The energy momentum for (Y-M) is,

$$\mathbf{T}_{\alpha\beta} = \langle \mathbf{F}^\mu_\alpha, \mathbf{F}_{\beta\mu} \rangle - \frac{1}{4} \mathbf{g}_{\alpha\beta} \langle \mathbf{F}_{\mu\nu}, \mathbf{F}^{\mu\nu} \rangle.$$

It satisfies the local conservation laws (1.1.) as well as the positivity condition[5] (1.2.)

Unlike the Maxwell equations which can be expressed entirely in terms of $\mathbf{F}$, the Yang-Mills equation depend implicitly on the connection $\mathbf{A}$. Thus a solution of the Yang-Mills equations is a *class of gauge equivalent* connection 1-forms $\mathbf{A}$ whose curvature $\mathbf{F}[\mathbf{A}]$ verifies (Y-M). In practice, when solving the Yang -Mills equations we need to impose additional conditions which fix this gauge ambiguity. In these lectures I shall refer to three classical gauge conditions, the Lorentz gauge condition

$$\partial_\alpha \mathbf{A}^\alpha = 0, \qquad\qquad\qquad (\text{L})$$

the Temporal gauge

$$A_0 = 0 \qquad\qquad\qquad (\text{T})$$

and the Coulomb gauge,

$$\text{div } A = 0. \qquad\qquad\qquad (\text{C})$$

where $\text{div } A = \nabla^i A_i$.

From the point of view of the standard classification of partial differential equations into elliptic, parabolic or hyperbolic types the most convenient of all is the Lorentz gauge. Indeed in that case the Yang-Mills equations take the form of a system of non-linear wave equations,

$$\Box_\mathbf{g} \mathbf{A}_\alpha = N_\alpha(\mathbf{A}, \partial\mathbf{A}) \qquad\qquad\qquad (1.3)$$

with N a non-linear term $N = Q + C$, where $Q = Q(\mathbf{A}, \partial\mathbf{A})$ is quadratic in $\mathbf{A}$ and its space-time derivatives[6] $\partial\mathbf{A}$, while C is a term cubic in $\mathbf{A}$. Despite the relative simplicity of the form of the Yang-Mills equations in the Lorentz gauge we shall see that the other two gauge conditions, which lead to systems which have a mixture of both hyperbolic and elliptic features, are in fact better behaved.

Example 4. The Einstein Field Equations
According to the general relativistic variational principle the space-time metric $\mathbf{g}$ itself is stationary with respect to an action $\mathcal{S} = \int_\mathcal{U} L dv_\mathbf{g}$ whose Lagrangean

---

5)   This requires the positivity of the Killing form of $\mathfrak{g}$.

6)   Also linear with respect to $\partial\mathbf{A}$.

$L = L_G + L_M$ consists of a purely gravitational part $L_G = \mathbf{R}$, given by the scalar curvature $\mathbf{R}$ of the metric $\mathbf{g}$, and the matter Lagrangean L_M which depends on the matter fields $\mathbf{H}$ present in the space-time. The corresponding Euler-Lagrange equations are the so called Einstein field equations,

$$\mathbf{R}_{\mu\nu} - \frac{1}{2}\mathbf{g}_{\mu\nu}\mathbf{R} = \mathbf{T}_{\mu\nu} \tag{E}$$

where $\mathbf{R}_{\mu\nu}$ is the Ricci curvature of $\mathbf{g}$ and $\mathbf{T}_{\mu\nu}$ the energy momentum tensor of matter, $\mathbf{T}_{\mu\nu} = \left(\frac{\partial L_M}{\partial \mathbf{g}^{\mu\nu}} - \frac{1}{2}\mathbf{g}_{\mu\nu}L_M\right)$ obtained from the variation of $S_M = \int L_M dv_{\mathbf{g}}$ with respect to the metric. Thus the energy momentum tensors of the matter fields discussed before arise from the variations of their action integrals with respect to the background metric. From this perspective we remark that the purely gravitational part of the Einstein field equations has no energy-momentum tensor, as defined for the other field theories, since the left hand side of the equations (E) have been derived precisely by varying the action $\mathcal{S}_G$ relative to $\mathbf{g}$. This formal difficulty is connected to the well known difficulty of principle of defining local concepts of energy, linear and angular momenta for General Relativity as we have them for the other field theories.

As is well known Einstein proposed General Relativity as a unified theory of space, time and gravity. In fact, according to it, gravity is a manifestation of the space-time. An important consequence of this is that the Einstein field equations (E) admit nontrivial solutions even in the absence of external matter fields. This amounts to setting $\mathbf{T} \equiv 0$ on the right-hand side of (E). The resulting equations, called the Einstein-Vacuum equations, take the form,

$$\mathbf{R}_{\alpha\beta} = 0. \tag{E-V}$$

Most of the fundamental difficulties of the general Einstein field equations are present in the form (E-V). From a physical point of view the equations (E-V) are perfectly adequate to describe the propagation of gravitational waves at large distances from a source.

Going back to the general case (E) we recall that the Riemann curvature tensor satisfies the Bianchi identities,

$$\mathbf{D}_\epsilon \mathbf{R}_{\alpha\beta\gamma\delta} + \mathbf{D}_\alpha \mathbf{R}_{\beta\epsilon\gamma\delta} + \mathbf{D}_\beta \mathbf{R}_{\epsilon\alpha\gamma\delta} = 0. \tag{B}$$

Contracting (B) twice we derive, $\mathbf{D}^\nu(\mathbf{R}_{\mu\nu} - \frac{1}{2}\mathbf{g}_{\mu\nu}\mathbf{R}) = 0$. This implies that the right-hand side of (E) also satisfies,

$$\mathbf{D}^\nu \mathbf{T}_{\mu\nu} = 0$$

which, remarkably, is precisely the local law of conservation for the energy-momentum tensor of the matter-field.

In view of the four contracted Bianchi identities equations (E-V) can be viewed as a system of $10 - 4 = 6$ equations for the 10 components of the metric tensor $\mathbf{g}$. The remaining 4 degrees of freedom correspond to the general covariance of the equations. Thus, if ϕ is a diffeomorphism of $\mathcal{M}$, then the pair $(\mathcal{M}, \mathbf{g})$ and $(\mathcal{M}, \phi_* \mathbf{g})$ represent the same solution of the field equations. Written explicitly in an arbitrary system of coordinates equations (E-V) lead to a nonlinear degenerate system of partial differential equations in $\mathbf{g}$. Indeed the principal part of the Ricci curvature $\mathbf{R}_{\mu\nu}$ is $\frac{1}{2}\, \mathbf{g}^{\alpha\beta}\Big(\partial_\mu\partial_\alpha \mathbf{g}_{\beta\nu} + \partial_\nu\partial_\alpha \mathbf{g}_{\beta\mu} - \partial_\mu\partial_\nu \mathbf{g}_{\alpha\beta} - \partial_\alpha\partial_\beta \mathbf{g}_{\mu\nu}\Big)$. Nevertheless the (E-V) equations are seen to be hyperbolic when one takes into account the geometric equivalence of metrics related by diffeomorphisms. This indicates that it must be possible to choose coordinates relative to which the (E-V) equations can be written as a system of nonlinear wave equations. One well known such choice is that of "wave coordinates" namely coordinates x^μ which satisfy the wave equation $\mathbf{g}^{\alpha\beta}D_\alpha D_\beta x^\mu = 0$. In this case the principal part of $\mathbf{R}_{\mu\nu}$ becomes $-\frac{1}{2}g^{\alpha\beta}\partial_\alpha\partial_\beta \mathbf{g}_{\mu\nu}$ and the (E-V) equations take the form:

$$\mathbf{g}^{\alpha\beta}\partial_\alpha\partial_\beta \mathbf{g}_{\mu\nu} = F_{\mu\nu}(\mathbf{g}, \partial\mathbf{g}) \tag{1.4}$$

with $F_{\mu\nu}$ a polynomial in $(\mathbf{g}, \partial\mathbf{g})$ quadratic in $\partial\mathbf{g}$. The form (1.4) allowed Y. Choquet-Bruhat [Br1] to prove the well posedness of the local Cauchy problem in wave coordinates by appealing to the local theory of systems of nonlinear wave equations developed by J. Leray. The reliance on fixed coordinate conditions is however very problematic if one wants to study the (E-V) equations in the large. Indeed, as pointed by Y. Choquet-Bruhat [Br2], the "wave coordinates" are unstable in the large even when one starts with almost flat initial conditions. This problem, often referred to as the problem of coordinates, is the first major difficulty one has to overcome in the construction of global solutions to the (E-V) equations.

Example 5. The reduced Einstein Vacuum Equations
This is the case of special solutions of the Einstein vacuum equations (E-V) which possess a space-like Killing vectorfield with no critical points. The reduced system can be expressed relative to a $2 + 1$ dimensional space-time $(\mathcal{M}, \mathbf{g})$ and a map $\phi : \mathcal{M} \longrightarrow H^2$ where H^2 is the 2-dimensional Poincaré plane. The map ϕ is supposed to be a Wave Map connected to the metric $\mathbf{g}$ by the equation

$$\mathbf{R}ic = \frac{1}{2}\phi^* h,$$

where $\mathbf{R}ic$ is the Ricci tensor of $\mathbf{g}$ and $\phi^* h$ is the pull-back of the Poincaré metric h.

The energy-momentum tensor of a field theory is intimately connected with Noether's Principle which asserts the following:

Noether's Principle. *To any one parameter group of transformation which preserves the action there corresponds a conservation law.*

To illustrate it let $\mathcal{S} = \mathcal{S}[\mathbf{H}, \mathbf{g}]$ be the action integral of the fields $\mathbf{H}$ and consider χ_t a 1-parameter group of isometries of $\mathcal{M}$, i.e. $(\chi_t)_* \mathbf{g} = \mathbf{g}$. Then

$\mathcal{S}[(\chi_t)_*\mathbf{H}, \mathbf{g}] = \mathcal{S}[(\chi_t)_*\mathbf{H}, (\chi_t)_*\mathbf{g}] = \mathcal{S}[\mathbf{H}, \mathbf{g}]$. Therefore, in view of Noether's Principle, to any 1- parameter group of isometries of the space-time there corresponds a conservation law. The standard way to find these conservation laws goes as follows:

First let X be an arbitrary vectorfield and $\mathbf{P}$ be the 1-form obtained by contracting the energy-momentum tensor of $\mathbf{H}$ with X i.e. $\mathbf{P}_\alpha = \mathbf{T}_{\alpha\beta}X^\beta$. Then, since $\mathbf{T}$ is symmetric and divergence-less,

$$D^\alpha \mathbf{P}_\alpha = \frac{1}{2}\mathbf{T}^{\alpha\beta}\,\pi_{\alpha\beta} \tag{1.5}$$

where,

$$\pi_{\alpha\beta} = D_\beta X_\alpha + D_\alpha X_\beta$$

is the deformation tensor of X. We now integrate (1.5) on a lens shaped domain $\mathcal{D}$ bounded by two space-like hypersurfaces $\mathcal{H}_0$, $\mathcal{H}_1$. In view of the divergence theorem we derive the integral identity,

$$\int_{\mathcal{H}_1} \mathbf{T}(X, T)da_g - \int_{\mathcal{H}_0} \mathbf{T}(X, T)da_g = -\int_{\mathcal{D}} \mathbf{T}^{\alpha\beta}\,\pi_{\alpha\beta} \tag{1.6}$$

where T is the future oriented unit normal, g the induced metric and da_g the area element of $\partial\mathcal{D}$.

In the particular case when X is a Killing[7] vectorfield its deformation tensor π vanishes identically and we derive, the conservation law,

$$\int_{\mathcal{H}_1} \mathbf{T}(X, T)da_g = \int_{\mathcal{H}_0} \mathbf{T}(X, T)da_g \tag{1.7}$$

We remark that the result (1.7) remains true if X is a conformal Killing vector field, i.e. it generates conformal isometries, and $\mathbf{T}$ is traceless. Indeed, if X is conformal Killing, $\pi_{\alpha\beta} = \Lambda \mathbf{g}_{\alpha\beta}$ and hence $\pi_{\alpha\beta}\mathbf{T}^{\alpha\beta} = \Lambda tr(\mathbf{T}) = 0$. One can easily show that if the action integral $\mathcal{S}[\mathbf{H}, \mathbf{g}]$ is invariant under conformal rescalings of the metric, i.e. $\tilde{\mathbf{g}} = \Omega^2\mathbf{g}$, then the corresponding energy-momentum tensor is traceless. This is the case of the Yang-Mills theory in $3 + 1$ dimensions.

The identities (1.6) and (1.7) are usually applied to time-like future oriented vectorfields X in which case, in view of the positive energy condition, the integrand $\mathbf{T}(X, T)$ is positive. In particular, in Minkowski space-time we can choose $X = \frac{\partial}{\partial t}$. Choosing also the hypersurfaces $\mathcal{H}_0$ to be the level surfaces of the time function t such that also $T = \frac{\partial}{\partial t}$, the formula (1.7) becomes the standard form of Conservation of Energy,

$$\mathcal{E}(t) = \mathcal{E}(0).$$

where $\mathcal{E}(t)$ is the integral of the quantity $\mathbf{T}_{00} = \mathbf{T}(T, T)$ along the t-hyperplanes.

Finally, to end this preliminary discussion I ought to add a few more words concerning the Initial Value Problem. It turns out that the general solutions of

---

7)   i.e. it generates a one parameter group of isometries

a given Field Theory can be best parametrized in terms of its initial data on a space-like hypersurface. Thus, for example, the general solutions to (N.W.E) can be specified by prescribing ϕ and its normal derivative on a fixed space-like hyperplane, e.g. $t = 0$ in Minkowski space-time. The same holds true for (W.M.) The situation for Gauge Theories such as the Yang-Mills and the Einstein equations is slightly more complicated.

Let $\mathcal{H}_0$ be a space-like hypersurface in our space-time $\mathcal{M}$, with T its future directed unit normal. If $\mathbf{A}$ is a Yang-Mills connection with curvature $\mathbf{F}$ we let A be the induced connection on $\mathcal{H}_0$. The curvature $\mathbf{F}$ decomposes into its electric part $E = i_T \mathbf{F}$ and magnetic part $H = i_T(*\mathbf{F})$. It is easy to see that H is in fact the curvature of the induced connection A. Thus an initial data set ought to be specified by A and E. However the Yang-Mills equations imply the additional constraint,

$$\mathrm{div}\, E + [A, E] = 0 \tag{1.8}$$

We define an initial data set on $\mathcal{H}_0$ to be a pair formed by a $\mathfrak{g}$ -valued connection A on $\mathcal{H}_0$ and a vector E which verify the equations (1.8.).

Similarly, in the case of the Einstein-Vacuum equations, assume that $\mathcal{H}_0$ is a spacelike hypersurface embedded in a given space-time $(\mathcal{M}, \mathbf{g})$ verifying (E-V). Let g, k be the first and second fundamental forms of $\mathcal{H}_0$ and denote by ∇ the induced covariant differentiation. Since $\mathcal{H}_0$ is space-like the metric g is Riemannian. Now the Gauss-Codazzi equations of the embedding of $\mathcal{H}_0$ in $\mathcal{M}$ together with the equations (E-V) lead to the following nonlinear system of equations satisfied by g and k in $\mathcal{H}_0$.

$$\nabla^j k_{ji} - \nabla_i \mathrm{tr} k = 0$$
$$R - |k|^2 + (\mathrm{tr} k)^2 = 0 \tag{1.9}$$

called the constraint equations. Here R is the scalar curvature of g, trk is the trace of k relative to g and $|k|^2 = k_{ij} k^{ij}$. In view of this we define an initial data set for (E-V) to be a triplet $(\mathcal{H}_0, g, k)$ formed by a Riemannian 3-D manifold $(\mathcal{H}_0, g)$ together with k, a symmetric tensor of rank 2, which verify the constraint equation (1.9.) A development of an initial data set consists of an Einstein-Vacuum space-time $(\mathcal{M}, \mathbf{g})$ together with an embedding $i \colon \mathcal{H}_0 \to \mathcal{M}$ such that g, k are the induced first and second fundamental forms of the embedding.

One can formulate the central question of the mathematical theory of General Relativity as that of studying the evolution of general initial data sets. Now this goal is too broad since without an appropriate asymptotic restriction the evolution of an arbitrary initial data set can be very bad. A reasonable physical restriction is to consider initial data sets which look flat outside a sufficiently large compact set of $\mathcal{H}_0$. The evolution of such initial data sets correspond to "isolated physical systems". These systems are particularly important in G.R. since, as was stressed by Geroch [Ge], "it is only through a suitable notion of an isolated system that one acquires an ability at all to deal individually with various subsystems in the universe; in particular to assign to subsystems such physical attributes as mass, angular momentum, character of emitted radiation, etc." Indeed, as we have mentioned above, the Einstein equations have no energy-momentum tensor and therefore no local concepts of energy, linear and angular momentum. Moreover for any field theory in which the metric is kept fixed these concepts arise from the symmetries of the background space-time. These do not make sense in G.R., general solutions have no symmetries. Global concepts of energy, linear and angular momentum are nevertheless possible when we restrict our attention to space-times with asymptotic symmetries at infinity. Such is the case of an isolated physical system; at large distances from the source the corresponding space-time becomes flat and thus acquires the symmetries of the Minkowski space-time.

An initial data set $(\mathcal{H}_0, g, k)$ is said to be asymptotically flat if there exists a system of coordinates x^1, x^2, x^3 globally defined, with a possible exception of a relatively compact set in $\mathcal{H}_0$ such that $g_{ij} - (1 + \frac{2M}{r})\delta_{ij} \to 0$, $k_{ij} \to 0$ as $r^2 = \Sigma(x^i)^2 \to \infty$. More precisely, we will say that the initial data set is strongly asymptotically flat, or S.A.F., of order[8] p if,

$$g_{ij} - (1 + \frac{2M}{r})\delta_{ij} = o_p(r^{-3/2}), \quad k_{ij} = o_{p-1}(r^{-3/2}) \qquad \text{(S.A.F.)}$$

Given an S.A.F. initial data set one can define its energy E, linear momentum P and angular momentum J by, $E = \frac{1}{16\pi} \lim_{r\to\infty} \int_{S_r} \sum_{i,j} \left(\partial_i g_{ij} - \partial_j g_{ii} \right) N^j da$, $P_i = \frac{1}{8}\pi \lim_{r\to\infty} \int_{S_r} (k_{ij} - \text{tr}k g_{ij}) N^j da$, $i = 1, 2, 3$ and, $J_i = \frac{1}{8}\pi \lim_{r\to\infty} \int_{S_r} \epsilon_{iab} x^a (k^{bj} - g^{bj} \text{tr}k) N_j da$, $i = 1, 2, 3$ where S_r is the coordinate sphere of radius r, N the exterior unit normal to it and da its area element. In view of (S.A.F) we also have $E = M$, $P = 0$. Thus the (S.A.F.) condition implies that the initial data set is in a center of mass frame. In view of the positive mass theorem [S-Y], M must be a positive number vanishing only if the initial data set is flat (i.e. its development is flat).

2. The Problem of Break-down

The break-down phenomenon can occur despite the existence of the basic conservation laws, in particular for the energy, or total mass, which is positive. To understand what this means consider the comparable situation in one dimension, namely systems of ordinary differential equations which arise as the Euler-Lagrange equations of a Lagrangean with positive energy. To be more precise consider the example of the differential equation,

$$\ddot{x} + V'(x) = 0$$

---

8)   A function f is said to be $o_m(r^{-p})$, resp. $O_m(r^{-p})$, as $r \to \infty$ if $\partial^l f(x) = o(r^{-p-l})$, resp. $O(r^{-l-p})$, for any $l = 0, 1 \ldots m$, where ∂^l denote all the partial derivatives of order l relative to the coordinates x^1, x^2, x^3.

subject to the initial conditions at $t = 0$,

$$x(0) = x_0, \quad \dot{x}(0) = x_1 .$$

The total energy of the system is given by the expression $\frac{1}{2}|\dot{x}|^2 + V(x)$, where $\frac{1}{2}|\dot{x}|^2$ is the kinetic energy and V the potential energy. For any reasonable physical system $V \geq 0$. Since the total energy is conserved we immediately conclude that, for any initial conditions, the corresponding solutions exist for all time. There can be no finite time blow-up of the solutions.

The situation is very different for classical field theories. Though we have a positive energy momentum tensor which leads, in Minkowski space-time, to well defined conserved quantities, we cannot infer in general that the solutions, starting with perfectly smooth initial conditions, remain so for all time. Take as an example the field theory closest to our one dimensional example, namely the scalar wave equation in Minkowski space-time $\mathbf{M}^{n+1}$,

$$\Box \phi - V'(\phi) = 0, \qquad\qquad (\text{N.W.E})$$

subject to the initial conditions at $t = 0$,

$$\phi(0) = f_0, \quad \partial_t \phi(0) = f_1 .$$

Assume that f_0, f_1 are as regular as we want, say $f_0, f_1 \in C_0^\infty(\mathbf{R}^n)$. Assume also that $V \geq 0$ so that the energy-momentum tensor satisfies the required positivity condition.

The total energy[9] at time t has the form,

$$\mathcal{E}(t) = \int_{\mathbf{R}^n} \left(\frac{1}{2}((\partial_t \phi)^2 + (\partial_1 \phi)^2 \cdots + (\partial_n \phi)^2) + V(\phi) \right) dx$$

Thus, according to the law of conservation of energy, if the initial data at time $t = 0$ are such that $\mathcal{E}(0)$ is bounded, we infer that $\mathcal{E}(t)$ is bounded for all time. Yet, unlike in the previous example we cannot conclude that the solutions remain smooth at all later times. It is easy to see that as long as ϕ remains bounded its time evolution preserves the regularity of the initial conditions in the L^2 norm . The problem, thus, is to show that ϕ stays bounded. In space dimension $n = 1$ we can conclude, from the boundedness of $\mathcal{E}(t)$ and a simple form of the Sobolev inequalities, that ϕ is point-wise bounded. Indeed, since $V \geq 0$, we infer that at any time t, $\frac{1}{2}\int_{\mathbf{R}^n} \left((\partial_t \phi)^2 + (\partial_1 \phi)^2 \cdots + (\partial_n \phi)^2 \right) dx \leq \mathcal{E}(0)$. On the other hand, according to the simplest version of the basic Sobolev inequalities, the sup-norm of a function in $\mathbf{R}^n$ can be bounded in terms of the square integrals of the sum of all its derivatives of order[10] $[\frac{n}{2}]$. Hence for $n = 1$,

$$\sup_x |\phi(t,x)| \leq c\mathcal{E}(0)^{1/2} .$$

_____

9)   which is the conserved quantity corresponding to the Killing vectorfield $T_0 = \partial_t$

10) $[\frac{n}{2}]$ is the smallest integer strictly greater than $\frac{n}{2}$.

The form of the Sobolev inequality we have used above just fails for $n = 2$. In fact we can only estimate $(\int |\phi(t, x)|^p)^{1/p}$ for any $p < \infty$, this turns out nevertheless to be enough to conclude that the solutions remain smooth for all time[11]. For $n \geq 3$ the above form of the Sobolev inequality require a little more than $\frac{3}{2}$ derivatives in L^2 in order to estimate the sup norm of ϕ. The boundedness of the energy provides us with a bound of only one derivative in L^2. We have thus a gap of more than a half derivative. The situation gets, of course, even worse in higher dimensions. We can still, nevertheless, show that ϕ remains bounded provided that $V(\phi)$ does not grow too fast as $\phi \rightarrow \infty$. For simplicity consider the case of the power potential $V(\phi) = \frac{1}{p+1}|\phi|^{p+1}$. Prescribing to the space-time variables $t = x^0, x^1, \ldots, x^n$ the same scale L, the solution ϕ of the equation, $\Box\phi - V'(\phi) = 0$, acquires the scale $L^{-\frac{2}{p-1}}$. Therefore the total energy $\mathcal{E}$ has the scale L^s where s is the exponent $s = n - 2 - \frac{4}{p-1}$. We shall call $\bar{s} = \frac{1}{2}(n - 2 - \frac{4}{p-1})$ the scaling exponent of our field equation.

The case when the exponent $\bar{s}$ is strictly negative is called "subcritical". It is quite easy to analyze and has lead to the well-known global regularity result of Jörgens [Jö]. The case $\bar{s} = 0$ is called "critical" while $\bar{s} > 0$ is called "supercritical." In the supercritical regime we have no results, even for spherically symmetric solutions, despite the relatively large attention this problem has received. The critical case has been recently settled by the combined efforts of Struwe [Stru], Grillakis [Gr1] and more recently Shatah-Struwe [Sh-Stru1].

Theorem 1. *Consider the initial value problem $\Box\phi - V'(\phi) = 0$ with initial conditions $\phi(0, x) = f_0(x), \partial_t\phi(0, x) = f_1(x)$ which, for simplicity, we may assume in C_0^∞. Assume that $n \leq 7$ and that the scaling exponent $\bar{s} \leq 0$. Then the equation admits unique smooth solutions globally in $\mathbf{M}^{n+1}$.*

A more interesting field theory is provided by the equations of Wave Maps defined from Minkowski space-time $\mathbf{M}^{n+1}$ with values in a Riemannian manifold $\mathcal{N}$. Relative to standard coordinates x^α, $\alpha = 0, \ldots, n$ and local coordinates y^a, $a = 1, \ldots m$ in $\mathcal{N}$ the equations take the form,

$$\Box\phi^a + \Gamma_{bc}^a(\phi)\mathbf{m}^{\mu\nu}\partial_\mu\phi^b\partial_\nu\phi^c = 0 \qquad \text{(W.M.)}$$

where Γ_{bc}^a are the Christoffel symbols of $\mathcal{N}$. Consider the initial conditions at $x^0 = t = 0$,

$$\phi(0) = f_0, \quad \partial_t\phi(0) = f_1$$

with f_0, f_1 compactly supported smooth maps defined from $\mathbf{R}^n$ to $\mathcal{N}$. Since the nonlinear terms are quadratic in the first derivatives of the map ϕ, in order to preserve the L^2 regularity of the initial conditions, we now need to have pointwise bounds not only on ϕ but also its first derivatives. The total energy, in this case, has the form,

$$\mathcal{E}(t) = \frac{1}{2}\int_{\mathbf{R}^n}\left(|\partial_t\phi|^2 + |\partial_1\phi|^2 \cdots + |\partial_n\phi|^2\right)dx$$

---

11)   provided that $V(\phi)$ has polynomial growth in ϕ for large ϕ.

with $|\partial_\alpha \phi|^2 = h_{ab}\partial_\alpha \phi^a \partial_\alpha \phi^b$, h the Riemannian metric of $\mathcal{N}$. The law of conservation of total energy is, as before,

$$\mathcal{E}(t) = \mathcal{E}(0) .$$

This provides us only with an L^2 bound for the derivatives of ϕ. We therefore see that the conservation law for the total energy does not suffice to control the L^∞ norm of the first derivatives of ϕ even for wave maps in $\mathbf{M}^{1+1}$. A simple remark, however, allows us to bypass the difficulty in this case (see [Sh] and also [Gu]). Indeed consider the energy momentum tensor $\mathbf{T}$. It has the form $\mathbf{T}_{\alpha\beta} = \frac{1}{2}\left(\langle \phi_{,\alpha}, \phi_{,\beta}\rangle - \frac{1}{2}\mathbf{g}_{\alpha\beta}(\mathbf{g}^{\mu\nu}\langle \phi_{,\mu}\phi_{,\nu}\rangle)\right)$ with $<,>$ the scalar product in $\mathcal{N}$. Since we are in $1+1$ dimensions $\mathbf{T}$ has only the components $\mathbf{T}_{00}, \mathbf{T}_{01} = \mathbf{T}_{10}, \mathbf{T}_{11}$. Moreover, since $\mathbf{T}$ is trace-less in $\mathbf{M}^{1+1}$, we have $\mathbf{T}_{00} = \mathbf{T}_{11}$. Now recall that $\mathbf{T}$ verifies the divergence equation $\partial^\beta \mathbf{T}_{\alpha\beta} = 0$. Hence,

$$\partial_t \mathbf{T}_{00} = \partial_x \mathbf{T}_{01}$$
$$\partial_t \mathbf{T}_{01} = \partial_x \mathbf{T}_{00}$$

and therefore $\mathbf{T}_{00}$ is a solution of the linear wave equation in $\mathbf{M}^{1+1}$, $\Box \mathbf{T}_{00} = 0$. One can now easily check that $\mathbf{T}$ remains bounded for any $t > 0$ provided that $\mathbf{T}_{00}, \partial_t \mathbf{T}_{00}$ are bounded at $t = 0$. Since $\mathbf{T}_{00} = \frac{1}{2}(|\partial_t \phi|^2 + |\partial_x \phi|^2)$ we conclude that all first derivatives of the map ϕ are bounded. Therefore, in $\mathbf{M}^{1+1}$, all wave maps which are smooth initially remain so for all time.

The proof we have presented is typical to the sweeping simplifications which occur only in $1+1$ dimensions. The case of wave maps defined in $\mathbf{M}^{1+2}$ is far more complicated. We can proceed as before, in the case of N.W.E. and classify the W.M. according to the scale associated to the total energy $\mathcal{E}$. Thus, prescribing to the space-time variables the scale L and to ϕ the scale L^0 we find that $\mathcal{E}$ has the scale L^s with $s = n-2$. As before we call $\bar{s} = \frac{n-2}{2}$ the scaling exponent of the field theory. Consequently the W.M. is subcritical in $\mathbf{M}^{1+1}$, critical in $\mathbf{M}^{1+2}$ and supercritical in $\mathbf{M}^{1+n}$, $n \geq 3$. Under reasonable geometric assumptions we expect global regularity in the critical case $\mathbf{M}^{1+2}$. This conjecture has been recently checked for wave maps satisfying additional symmetry assumptions, see [Ch-Za] in the case of spherical symmetry and [Sh-Za2], [Gr2] in the equivariant case. However the general case remains wide open. On the other hand, in the supercritical case $n \geq 3$, singularities can in fact occur. We owe to Shatah [Sh] and Shatah-Tahvildar-Zadeh [Sh-Za1] a simple example of selfsimilar equivariant wave map from $\mathbb{R}^{3+1}$ to S^m which is perfectly smooth for $t \leq 1$ but breaks down precisely at $t = 1$.

The Yang-Mills equation can also be tested from the point of view of the scaling analysis we have discussed above. Once more, if we prescribe the scale L to the space-time variables, then the gauge potential $\mathbf{A}$ scales like L^{-1} and its curvature $\mathbf{F}$ like L^{-2}. Thus the total energy at given t has the scale L^{-4+n}. The scaling exponent is $\bar{s} = \frac{-4+n}{2}$. Therefore the Yang-Mills equation are subcritical in the physical dimension $n = 3$, critical for $n = 4$ and supercritical for $n \geq$

4. The problem of global regularity for the Yang-Mills equation in Minkowski space-time, for sufficiently regular initial data, was solved in a beautiful paper by Eardley-Moncrief [E-M]. The proof required an insightful observation concerning the structure of the nonlinear terms of the Yang-Mills equations expressed in the Cronström gauge. It also uses in an essential way the explicit form of the fundamental solution of the wave equation in the flat Minkowski space-time.

Finally, in view of the same simple minded scaling analysis we can easily check that, relative to the total ADM mass[12], the Einstein field equations are supercritical[13]. On the other hand the reduced Einstein equations are critical.

General Conjecture.
(i)   *The interesting[14] subcritical field theories are regular for all smooth data[15].*
(ii)   *Under reasonable restrictions the critical field theories are regular for all smooth data.*
(iii) *"Sufficiently small" solutions to the supercritical field theories are regular. The general, large, solutions develop singularities in finite time.*

The first two parts of the conjecture have been, so far, verified only for the critical and subcritical scalar wave equation, for the Yang-Mills equations in $\mathbb{R}^{3+1}$, and for the spherically symmetric and equivariant Wave Maps equations with certain geometric assumptions of the target manifold. It is believed by many that the scalar nonlinear wave equations are regular even in the supercritical cases. This is based on the fact that numerical calculations fail to produce large amplitudes. It is however possible that the break-down phenomenon is unstable and thus impossible to detect by numerical calculations. The regularity of solutions for small initial conditions has been verified in the case of supercritical Wave Maps[16] and in the case of the Einstein vacuum equations.

In these Lectures I will present our attempts to attack the Conjecture, or rather the regularity part of it, from the local point of view the "well posedness" of the initial value problem. According to this, if we can establish that the I.V.P. is well posed in the energy norm[17] then global regularity is automatically implied by Energy conservation.

---

12)   The total ADM mass is the positive conserved quantity for the Einstein equations playing the role of total energy.

13)   We assign to the space-time metric the scale L^0 and remark that the ADM mass $E = \frac{1}{16\pi} \lim_{r \to \infty} \int_{S_r} \sum_{i,j} \left(\partial_i g_{ij} - \partial_j g_{ii} \right) N^j da$ has the "supercritical" scale L^1.

14)   What I mean by interesting may be difficult to define. It should include most well known examples of relativistic field theories such as the ones discussed above. In what follows I will omit the word.

15)   This means that solutions which are smooth in the past cannot become singular.

16)   see discussion below

17)   This means that we have local existence, uniqueness and continuous dependence of the data in the energy norm. See also the more precise definition given below.

Conjecture 1. *For all subcritical field theories and, in an appropriate sense, for critical field theories the I.V. P. is well posed in the energy norm.*

The problem of well posedness is that of local in time existence, uniqueness and continuous dependence for the development of an initial data set. The classic results, based on energy estimates and Sobolev inequalities, apply to very large classes of non-linear wave equations but require too much differentiability on the initial data. Take, for example, the case of the Wave Maps equations[18]. Consider initial data sets (ϕ_0, ϕ_1) such that,

$$\|(\phi_0, \phi_1)\|_{H^s(\mathbf{R}^n)} = \|\phi(0, \cdot)\|_{H^{s+1}(\mathbf{R}^n)} + \|\partial_t\phi(0, \cdot)\|_{H^s(\mathbf{R}^n)} \leq \Delta. \qquad (2.1)$$

The classical local existence result reads as follows:

Proposition 2.1 *Let s_0 be a fixed exponent $> \frac{n}{2}$. There exists a time $T > 0$, depending only on s_0 and Δ, such that any initial data (ϕ_0, ϕ_1) verifying (2.1) with $s = s_0$ admit a unique development defined in a time slab $[0, T] \times \mathbf{R}^n$ which verifies, for all $0 \leq t \leq T$, and $s \geq s_0$,*

$$\|\phi(t, \cdot)\|_{H^{s+1}(\mathbf{R}^n)} + \|\partial_t\phi(t, \cdot)\|_{H^s(\mathbf{R}^n)} \leq C\left(\|\phi(0, \cdot)\|_{H^{s+1}(\mathbf{R}^n)} + \|\partial_t\phi(0, \cdot)\|_{H^s(\mathbf{R}^n)}\right)$$
$$(2.2)$$

Definition. *In what follows we will refer to a result of the type discussed in Proposition 2.1. by stating that the I.V.P is well posed in H^{1+s_0}.*

According to the Proposition the solution can be extended as long as the norm

$$\sup_{0 \leq t \leq T} \|\phi(t, \cdot)\|_{H^{s_0+1}(\mathbf{R}^n)} + \|\partial_t\phi(t, \cdot)\|_{H^{s_0}(\mathbf{R}^n)} \qquad (2.3)$$

remains finite. Moreover the solution preserves the H^s regularity of the initial conditions for all $s \geq s_0$. Let $T^* = T^*(s_0, \Delta)$ be the supremum of the values of T for which the norm in (2.3) remains finite. This is called the life-span of solutions corresponding to all initial data verifying (2.1.) The Proposition 2.1 asserts in fact that,

$$T^*(s, \Delta) > 0 \qquad (2.4)$$

for all $s > \frac{n}{2}$ and $\Delta > 0$.

Similar results hold for the Yang-Mills and Einstein-Vacuum equations. For example, in the case of the Yang-Mills equations in $\mathbf{M}^{1+3}$ relative to the Lorentz gauge, the classical result of local existence and uniqueness requires that the initial data set (A, E) be in $H^{s_0}(\mathbf{R}^3)$[19], i.e.

$$\|(A, E)\|_{H^{s_0}(\mathbf{R}^3)} = \|A\|_{H^{s_0+1}(\mathbf{R}^3)} + \|E\|_{H^{s_0}(\mathbf{R}^3)} \leq \Delta, \qquad (2.5)$$

with $s_0 > \frac{1}{2}$.

---

18) For simplicity throughout the discussion we shall assume that the entire manifold $\mathcal{N}$ is covered by one coordinate chart.

19) In fact it suffices to work with local H^s spaces.

For the Einstein-Vacuum equation the classical local existence result, obtained by Y. Chocquet-Bruhat in wave coordinates, requires the metric g to belong, locally, to H^{s_0} and the second fundamental form k to H^{s_0-1}, $s_0 > \frac{5}{2}$.

In all the above examples the classical local existence result, which is based only on energy estimates and Sobolev inequalities, requires one more derivative[20] than what we hope to be the optimal result. The statement below is a more general and version of Conjecture 1.

Conjecture 2. *For the field theories of scaling exponent $\bar{s}$ the initial value problem is, in an appropriate sense, locally well posed in the Sobolev spaces H^{s+1}, $s > \bar{s}$. Moreover, in a weak sense, the I.V.P is well posed in $H^{\bar{s}+1}$.*

Let us interpret Conjectures 1 and 2 in the case of the Wave Maps equations. Define the number,

$$s_c = \inf\{s \in \mathbf{R}, \text{ such that there exists } \Delta > 0 \text{ for which} T^*(s, \Delta) > 0. \ \} \quad (2.6)$$

To prove Conjecture 2 one would have to verify that this "optimal regularity exponent" is equal to the scaling exponent defined before, i.e.

$$s_c = \bar{s} = \frac{n-2}{2}$$

and that,

$$T^*(\frac{n-2}{2}, \Delta) > 0. \quad (2.7)$$

However the statement of Proposition 2.1 would have to be modified in the critical case

$$s = s_c = \bar{s} = \frac{n-2}{2}.$$

To see this consider the blow-up example provided by Shatah [Sh]. Indeed let $\phi(t, x)$ be the solution, constructed by Shatah, which breaks down at $t = T^*$. Then, in view of the scaling properties of (W.M.), the maps $\phi_\lambda(t, x) = \phi(\lambda t, \lambda x)$ are also solutions of (W.M). It is easy to check that for all $\lambda \geq 1$,

$$\|\phi_\lambda(0, \cdot)\|_{H^{\frac{n}{2}}(\mathbf{R}^n)} + \|\partial_t \phi_\lambda(0, \cdot)\|_{H^{\frac{n-2}{2}}(\mathbf{R}^n)} \leq C.$$

with C a positive constant independent of λ. On the other hand, each ϕ_λ breaks-down at precisely $T^*_\lambda = \frac{1}{\lambda}T^*$. Thus for large λ the life span of the solutions, corresponding to a set of initial data bounded in the $H^{\frac{n-2}{2}}$ norm, tends to zero. This shows that in fact (2.2) cannot hold for all $\Delta > 0$. For large Δ we can still have local, and even global solutions, without the estimate[21] (2.2.)

---

20)   In fact it is slightly more than a derivative.

21)   See the results of Shatah-Struwe [Sh-Stru2] and Kapitanski [Ka2] for the critical nonlinear scalar wave equation.

A precise formulation of Conjectures 1 and 2 should contain a correct formulation of well posedness in the critical case $s = s_c$. The formulation should also take into account the geometric properties of the equations, indeed without a proper geometric interpretation of the Sobolev spaces H^s the statements of Conjectures 1 and 2 are meaningless[22].

According to Conjecture 2 the optimal local existence result for the Yang-Mills equations requires data (A, E) for which the norm $\|(A, E)\|_{H^s(\mathbf{R}^3)} < \infty$, $s \geq -\frac{1}{2}$, defined in (2.5), is finite. In the case of the Einstein field equations the optimal local existence result should be expressed in terms of a norm defined in terms of some Sobolev type norm on (g, k) involving no more than $\frac{3}{2}$ derivatives of g and $\frac{1}{2}$ derivatives of k.

As we have discussed above the classical local existence theorem requires one more derivative than Conjecture 2. Gaining back this derivative is an extremely challenging task requiring, on one hand, the development of new analytic methods and, on the other hand, a deep understanding of the geometric structure of the nonlinear field theories. However one may ask, in the supercritical case, whether Conjecture 2 has anything to do with the general problem of break-down. Leaving aside the development of new methods which, I believe, are and will be generated in the process of gaining the above mentioned derivative, Conjecture 2 is also connected with the regularity of small data for supercritical field theories. Here is in fact the following more precise reformulation of part (iii) of the General Conjecture.

Conjecture 3. *For supercritical field theories "small data" have globally regular and "asymptotically free" developments.*

By small data we mean smallness of an appropriate global, weighted , Sobolev norm. By asymptotically free we mean that the corresponding solutions behave, for large time, like solutions of the underlying linear problem. The Conjecture states the fact that the developments corresponding to small perturbations of the trivial initial data set remain close, in an appropriate sense, to the trivial solution. We state below two results of this type. The first concerns the Wave Maps equations:

Theorem 2. *Consider the Minkowski space-time $\mathbf{M}^{1+n}$, $n \geq 2$, and the initial smooth maps f_0, f_1 defined form $\mathbf{R}^n$ with values in a sufficiently small neighborhood of a point p in the target manifold $\mathcal{N}$. There exists a global smooth map $\phi : \mathbf{M}^{1+n} \longrightarrow \mathcal{N}$, verifying the initial conditions $\phi(0, \cdot) = f_0$, $\partial_t \phi(0, \cdot) = f_1$. Moreover the image of the map ϕ remains concentrated in a small neighborhood of p.*

The result was originally proved by Kovalyov [Ko]. It was later extended by Sideris [Si] to the case of maps which remain close to a geodesic in $\mathcal{N}$.

---

22)   This is the case of the geometric theories such as Wave Maps, Yang Mills, Einstein equations etc.

In the case of the Einstein equations Conjecture 3 is precisely the problem of stability of the Minkowski space-time. Namely the Minkowski space-time $\mathbf{M}^{3+1}$ is a special solution of (E-V) free of singularities. If an initial data set $\mathcal{H}_0, g, k$ is flat[23] its development is precisely $\mathbf{M}^{3+1}$. It is thus natural to ask what happens to the developments of initial data sets which are small perturbations of a flat initial data set. This problem is important from two points of view. First, all attempts[24] to find explicit, non-flat, solutions of (E-V) equations, in the asymptotically flat regime, have lead to singular space-times. Second, any asymptotically flat initial data set can be interpreted, outside a sufficiently large relatively compact set $\mathcal{K}$, as a small perturbation of the flat initial data set. Thus the study of the global stability of Minkowski space-time is also the study of the asymptotic properties of the development of any asymptotically flat initial data set outside the future set of a sufficiently large set $\mathcal{K} \subset \mathcal{H}$.

The problem of the stability of Minkowski space-time has been recently addressed in my joint work with D. Christodoulou [Ch-Kl2]. The result which we were able to prove asserts the following,

Theorem 3 [Ch-Kl2]. *Any S.A.F.[25] initial data set which satisfies, in addition, a Global Smallness Assumption, leads to a unique, smooth solution of the Einstein-Vacuum equations, which is a geodesically complete development of the initial data. Moreover, this development is globally asymptotically flat, by which we mean that its Riemann curvature tensor approaches zero[26] on any causal or space-like geodesic, as the corresponding affine parameter tends to infinity.*

The global smallness assumption requires that an appropriate weighted L^2-norm of up to 2 derivatives of the curvature tensor of g and 3 derivatives of k are small. The smallness assumption in Theorem 2 requires an appropriate weighted L^2-norm of more than $n + 1$ derivatives of the initial data. Both results are clearly not optimal. We expect that the optimal result should require no more derivatives than needed in the problem of local well posedness. This is the content of the following stronger version of Conjecture 3.

Conjecture 4. *There exists an appropriate global L^2 norm, involving the minimum number of derivatives of the data for which the I.V.P. is locally well posed, whose smallness implies global regularity and some weak version of "asymptotic freedom".*

---

23)   In other words g is the euclidean metric and $k = 0$.

24)   In particular this is the case for the two parameter family of stationary solutions, called the Kerr family, which are all singular with the exception of the trivial member of the family – the Minkowski space-time.

25)   We note that the precise fall-off conditions of the initial data set are in fact given in L^2-weighted norms.

26)   Our result gives precise information on the rate of decay of different components of the curvature tensor.

Of course we expect that if the "smallness condition" is not satisfied solutions could break-down. In the case of Wave Maps we do have the singularity results of Shatah [Sh]. However we are very far from understanding how the singularities form in general. In the case of General Relativity we are still in a very primitive stage of understanding how black holes and singularities form. We have the famous incompleteness theorem of Penrose which asserts that if an initial data set of a space-time verifying the Einstein field equations (with very general assumptions on the energy-momentum tensor $\mathbf{T}$) has a trapped sphere[27] then some outgoing null geodesics normal to S must be future-incomplete. It is nevertheless not at all clear how trapped surfaces form, if they form at all from regular initial conditions, except in spherical symmetry situations. Unfortunately, however, the (E-V) equations do not allow interesting asymptotically flat solutions which are spherically symmetric[28] Thus at the present time we have no results concerning the formation of black holes for the (E-V) equations. On the other hand there are non-trivial spherically symmetric solutions for the Einstein field equations coupled with an additional matter field, such as a scalar field. The program of analyzing the general spherically symmetric solutions of the coupled Einstein-scalar wave equation was initiated and carried out with remarkable success by D. Christodoulou, see [Ch2], [Ch3].

The analogue of Conjecture 3 in the case of subcritical and critical field theories is,

Conjecture 5. *For critical and subcritical field theories all initial data, with regular behavior at space-like infinity, behave asymptotically free.*

Finally the question of analyzing the global regularity features of the developments of arbitrary initial dat set is a distant goal even in the case of the simplest supercritical Field Theories. In the case of General Relativity this is essentially the problem of the so called "Cosmic Censorship" conjecture of Penrose. Loosely speaking the conjecture asserts that, generically, there are no singularities outside black holes or, in in other words, there exist no stable naked singularities.

In the remaining sections of this paper I will present some recent results obtained in collaboration with M. Machedon concerning the problem of well posedness. Our results are part of an ongoing program to try to settle Conjectures 1 and 2 for the case of the Yang-Mills and Wave Maps equations. The main results we will discuss are:

Theorem 4. *Under some technical restrictions which will be discussed in section 4 the initial value problem for the Wave Maps equations in $\mathbb{R}^{n+1}$ is locally well posed in the space $H^{\frac{n}{2}+\epsilon}$.*

---

27) i.e. a space-like sphere S on $\mathcal{H}$ with a compact filling such that the outgoing null normals to S are everywhere converging

28) Except, of course, for the Schwarzschild space-time itself.

Theorem 5 [Kl-Ma2]. *Any locally H^1 finite energy initial data set in $\mathbf{R}^3$ admits a unique, global, admissible, generalized[29] development[30] in the temporal gauge. Moreover, the admissible solutions preserve any additional H^s regularity that the initial data may have.*

In particular Theorem 5 implies the global regularity result of Eardley-Moncrief [E-M].

In section 3 below I shall give a discussion of the main new techniques needed in the proof of Theorem 5. The point of departure here remains the classical energy estimates. The novelty consists in isolating the quadratic terms involving derivatives in the nonlinear terms of the Yang-Mills equations and observing that, *in the Coulomb gauge*, there are subtle cancellations which can be taken into account in the L^2-space-time norm. These leads us to the "Null estimates" presented in Proposition 3.3. as well as Proposition 3.5. We first show how these estimates help to improve the optimal "well posed" exponent for a general class of systems, see (3.9.), verifying the "Null Condition". I then show how to use them for the Yang-Mills equations. In section 4, I make a brief presentation of the main ideas in the proof of the null estimates. Finally in section 5, I will indicate the proof of Theorem 4. This requires a somewhat different point of view than that of section 4 by circumventing the classical energy inequalities alltogether and working instead directly in space-time norms.

3. Energy estimates and the Problem of Optimal Local Well Posedness

In this section I will present the main estimates needed in the proof of Theorem 5. These estimates are to be combined with the classical energy estimates. Let us start by considering general systems of wave equations of the form:

$$\Box\phi = N(\phi, \partial\phi) \tag{3.1}$$

We consider two cases:

(I) $\qquad\qquad\qquad N(\phi, \partial\phi) = \phi \cdot \partial\phi + \text{ cubic } (\phi)$

(II) $\qquad\qquad\qquad N(\phi, \partial\phi) = \Gamma(\phi)\partial\phi \cdot \partial\phi$

---

29) A generalized solution of the Yang-Mills equations is defined as a class of gauge equivalent connection 1-forms for which there exists a sufficiently regular representative $\mathbf{A}$ with a well defined, locally integrable, curvature $\mathbf{F}[\mathbf{A}]$ which verifies (Y-M) in the sense of distributions.

30) We say that a generalized solution A of the Yang-Mills equation, in the temporal gauge, is admissible if

$$A(t, \cdot) \in C^0\left([0, T]; H^1_{\text{loc}}(\mathbf{R}^3)\right) \bigcap C^1\left([0, T]; L^2_{\text{loc}}(\mathbf{R}^3)\right)$$

and it can be approximated, in the corresponding topology, by smooth solutions.

The type (I) equations can be viewed as a caricature of the Yang-Mills equations. Recall, see equation (1.3), that (Y-M) take in fact the form (I) in the Lorentz gauge. The type (II) equation can be viewed as a general class of nonlinear wave equations resembling those satisfied by Wave Maps. The classical local existence result requires data $\phi(0,\cdot) \in H^{s_0+1}$, $\partial_t\phi(0,\cdot) \in H^{s_0}$ where $s_0 > \frac{n}{2}$ for equations of type (II) and $s_0 > \frac{n}{2} - 1$ for equations of type (I). The proof of these results rests on the standard energy inequality for solutions to the inhomogeneous wave equation,

$$\Box\phi = F. \tag{3.2}$$

We have,

$$\|\partial\phi(t,\cdot)\|_{L^2(\mathbb{R}^n)} \leq \|\partial\phi(0,\cdot)\|_{L^2(\mathbb{R}^n)} + \int_0^t \|F(t',\cdot)\|_{L^2(\mathbb{R}^n)}dt' \tag{3.3}$$

and, more generally for Sobolev norms, $\|\ \|_s = \|\ \|_{H^s(\mathbb{R}^n)}$

$$\|\partial\phi(t,\cdot)\|_s \leq \|\partial\phi(0,\cdot)\|_s + \int_0^t \|F(t',\cdot)\|_s dt'. \tag{3.4}$$

Applying the inequality (3.4) to the systems of type (II) we infer that, as long as $|\phi|_\infty$ remains bounded, and for any $s \geq 0$,

$$\|\partial\phi(t,\cdot)\|_s \leq C_s\|\partial\phi(0,\cdot)\|_s \exp\left(\int_0^t |\partial\phi(t',\cdot)|_\infty dt'\right). \tag{3.5}$$

In the standard proof for local existence, one uses the Sobolev inequality, $|\partial\phi(t',\cdot)|_\infty \leq C\|\partial\phi(t',\cdot)\|_{\frac{n}{2}+\epsilon}$ for any $\epsilon > 0$. Thus, very crudely,

$$\int_0^t |D\phi(t',\cdot)|_\infty dt' \leq Ct \sup_{0\leq t'\leq t} \|\partial\phi(t',\cdot)\|_{\frac{n}{2}+\epsilon}. \tag{3.6}$$

Combining the estimates (3.5) with (3.6) we easily derive the classical local existence theorem stated in Proposition 2.1. In the case of equations of type (I) the same argument proves a local existence result requiring one less derivative of the initial data. More precisely,

Theorem 3.1. *For equations of type (I) the I.V.P. is "well posed"*[31] *in $H^{\frac{n}{2}+\epsilon}$ while for equations of type (II) the I.V.P. is well posed in $H^{\frac{n}{2}+1+\epsilon}$.*

Can we do better? The energy estimates (3.3–3.6) seem solid[32], indeed it is well known that (3.3) is the only available inequality, for $n \geq 2$, in which the first derivatives of ϕ are estimated in terms of F alone, without loss of derivatives. This

---

31) Recall that this requires data $\phi(0,\cdot) \in H^{\frac{n}{2}+\epsilon}, \partial_t\phi(0,\cdot) \in H^{\frac{n}{2}-1+\epsilon}$.

32) see however Remark 2 below.

fact is crucial for us, since the nonlinear terms contain derivatives. On the other hand the estimate (3.6) is clearly wasteful. We lose a lot of information by giving up the time integration on the left hand side. To do better we shall make use the following estimate, for solutions of (3.2.).

Proposition 3.1. *Consider the inhomogeneous wave equation (3.2) in $\mathbb{R}^{n+1}$. Consider first the case $n \geq 3$. For every $\epsilon > 0$ there exists a constant C for which we have the estimate,*

(i) $\quad \left(\int_0^t |\phi(t', \cdot)|_\infty^2 dt' \right)^{1/2} \leq C \left(\|D\phi(0, \cdot)\|_{\frac{n-3}{2}+\epsilon} + \int_0^t \|F(t', \cdot)\|_{\frac{n-3}{2}+\epsilon} dt' \right).$

For $n = 2$ we have on the other hand[33]

(ii) $\quad \left(\int_0^t |D^{1/4}\phi(t', \cdot)|_\infty^4 dt' \right)^{1/4} \leq C \left(\|D\phi(0, \cdot)\|_\epsilon + \int_0^t \|F(t', \cdot)\|_\epsilon dt' \right).$

The proof of Proposition 3.1 , for $n \geq 3$ is an immediate consequence of the Sobolev inequalities and the following form of the Strichartz-Brenner inequality (see [P]),

Proposition 3.2. *Consider the inhomogeneous wave equation (3.2.) in $\mathbb{R}^{n+1}$, $n \geq 3$. For every $\epsilon > 0$ there exists a constant C for which we have the estimate,*

$$\left(\int_0^t \|D^{\frac{n-3}{2(n-1)}} \phi(t', \cdot)\|_{\frac{2(n-1)}{n-3}}^2 dt' \right)^{1/2} \leq c \|D\phi(0, \cdot)\|_\epsilon$$

Remark 1. *We remark that for $n = 3$ the sharp form of the inequality stated in Proposition 3.1 is valid provided that ϕ is spherically symmetric. Indeed in that case we have,*

$$\left(\int_0^t |\phi(t', \cdot)|_\infty^2 dt' \right)^{1/2} \leq c \left(\|D\phi(0, \cdot)\| + \int_0^t \|F(t', \cdot)\| dt' \right)$$

This is a straightforward application of the Hardy-Littlewood maximal function inequality applied to the specific form of the solution. This result is probably true also for any $n \geq 3$. The inequality fails in general however. This has been shown in [Kl-Ma1] in the case of dimension $n = 3$.

Proposition 3.2 allows us to prove the following sharper version of the local existence theorem (see [Po-Si] also [Kl-Ma3]).

Theorem 3.2. *For general systems of type (I) the I.V.P. is well posed in*

$$H^{\frac{n-1}{2}+\epsilon} \qquad \text{for } n \geq 3$$

$$H^{\frac{3}{4}+\epsilon} \qquad \text{for } n = 2$$

---

33)   For a proof of the inequality (ii) below see [P], [G-V], [Ka1], [L-S].

In the case of systems of type (II) the I.V.P. is well posed in

$$H^{\frac{n+1}{2}+\epsilon} \qquad \text{for } n \geq 3$$
$$H^{\frac{7}{4}+\epsilon} \qquad \text{for } n = 2$$

In the case of dimension $n = 3$ the results of Theorem 3.2 are sharp[34], in general. Indeed H. Lindblad [L] has proved that the particular case of an equation of type (I)

$$\Box\phi = \phi\partial_t\phi$$

and respectively equation of type (II)

$$\Box\phi = (\partial_t\phi)^2$$

is not well posed[35] in H^1, respectively H^2.

In view of the negative results of Lindblad we have to give up the full generality of systems (I), (II). To see what additional restrictions we should consider we have to investigate more closely the actual structure of the field theories we are interested in.

To start with consider the local coordinate form of the Wave Maps equations in Minkowski space:

$$\Box_g\phi^a + \Gamma^a_{bc}(\phi)\mathbf{m}^{\mu\nu}\partial_\mu\phi^b\partial_\nu\phi^c = 0 \tag{3.7}$$

Observe that the most dangerous term, quadratic in the first derivatives, $\mathbf{m}^{\mu\nu}\partial_\mu\phi^b\partial_\nu\phi^c$ is special. We have, $\mathbf{m}^{\mu\nu}\partial_\mu\phi^b\partial_\nu\phi^c = Q_0(\phi^b, \phi^c)$ where Q_0 is the null quadratic form (see [Kl1], [Kl2]),

$$Q_0(\phi, \psi) = \mathbf{m}^{\mu\nu}\partial_\mu\phi\partial_\nu\psi = -\partial_t\phi\partial_t\psi + \sum_i \partial_i\phi\partial_i\psi \tag{3.8a}$$

The other null quadratic forms introduced in [Kl1], [Ch1], [Kl2] were,

$$Q_{\alpha\beta}(\phi, \psi) = \partial_\alpha\phi\partial_\beta\psi - \partial_\beta\phi\partial_\alpha\psi \tag{3.8b}$$

Suppose we consider equations of type (II) where we only allow quadratic interactions of the type (3.8a), (3.8b). Would that make a difference concerning well-posedness? The question was raised by M. Machedon and I in [Kl-Ma1] where we proved the following result:

Theorem 3.3. *Consider systems of equations of type (II) of the form,*

$$\Box\phi^I + \sum_{J,K} \Gamma^I_{J,K}(\phi)B^I_{JK}(\partial\phi^J, \partial\phi^K) \tag{3.9}$$

---

34)   In dimension $n = 2$ the result is also sharp. However in higher dimension we expect that the sharp result, for general systems, can be lowered by an additional $\frac{1}{4}$ in dimension $n = 4$ and by $\frac{1}{2}$ for $n \geq 5$.

35)   They are however well posed in these spaces for spherically symmetric solutions.

where the B^I_{JK} are any of the null forms (3.8a–3.8b). For such systems the I.V.P is well posed in $H^{\frac{n+1}{2}}$.

The key new estimate which allows us to prove Theorem 3.2 is given by the following:

Proposition 3.3. *Consider the solutions ϕ, ψ of inhomogeneous wave equations in Minkowski space-time $\mathbb{R}^{n+1}$,*

$$\Box\phi = F, \qquad \Box\psi = G$$

with initial conditions,

$$\phi(0,x) = f_0(x), \quad \partial_t\phi(0,x) = f_1(x)$$
$$\psi(0,x) = g_0(x), \quad \partial_t\psi(0,x) = g_1(x).$$

Then, for any of the null forms (3.8a)–(3.8b) we have,

$$\int_0^T \int_{\mathbb{R}^n} |Q(\phi,\psi)|^2 \, dx dt$$

$$\leq c\Big(\|\nabla^{\frac{n+1}{2}} f_0\|_{L^2(\mathbb{R}^n)} + \|\nabla^{\frac{n-1}{2}} f_1\|_{L^2(\mathbb{R}^n)} + \int_0^T \|\nabla^{\frac{n-1}{2}} F(t,\cdot)\|_{L^2(\mathbb{R}^n)} dt\Big)^2 \quad (3.10)$$

$$\cdot \Big(\|\nabla g_0\|_{L^2(\mathbb{R}^n)} + \|g_1\|_{L^2(\mathbb{R}^n)} + \int_0^T \|G(t,\cdot)\|_{L^2(\mathbb{R}^n)} dt\Big)^2$$

Moreover, in dimensions $n = 2, 3$ the estimate is false if we replace the null forms (3.8a)-(3.8b) by arbitrary quadratic forms.

Remark 2. *The estimate (3.10) can be used in conjunction with the energy estimates (3.4) to prove Theorem 3.3. The method has however an obvious flaw. This is due to the inconsistency between the norm $\int_0^T \|F(t,\cdot)\|_{L^2(R^n)} dt$ which appears in the energy inequality (3.3) and the norm $\big(\int_0^T \|F(t,\cdot)\|^2_{L^2(R^n)} dt\big)^{\frac{1}{2}}$ needed to apply (3.10). Thus, while the method provides better results than the Strichartz-Brenner inequalities, we cannot expect it to yield the optimal results expected, according to Conjecture 2. To overcome this difficulty we will have to give up the energy inequalities (3.3-3.5) and consider instead the L^2 space-time framework suggested by (3.10). We shall discuss this appproach in section 5.*

It is more difficulty to see what is the special structure of the Yang-Mills equations which allows one to improve the general results of Theorem 3.1 for equations of type (I). To do that we need to use the gauge covariance of the equations to our advantage. Thus, while relative to the Lorentz gauge the Yang-Mills equations don't seem to exhibit any special structure, we shall see that the Coulomb gauge allows us to recast the equations in a form similar to (3.9). For convenience we rewrite the equations (Y-M) in the form,

$$\partial^\mu \mathbf{F}_{\alpha\mu} = \mathbf{J}_\alpha. \qquad \text{(Y-M')}$$

where
$$\mathbf{J}_\alpha = -\left[\mathbf{A}^\mu, \mathbf{F}_{\alpha\mu}\right].$$

Differentiating (Y-M') we notice that,

$$\partial^\alpha \mathbf{J}_\alpha = 0.$$

which is the law of conservation of charge.

In view of the Coulomb gauge condition

$$\nabla^i A_i = 0,$$

setting $\alpha = 0$ in (Y-M'), we derive

$$J_0 = \partial^i \mathbf{F}_{0i} = -\Delta A_0 + \partial^i \left[A_0, A_i\right]$$

Hence, since $J_0 = -\left[A_i, E_i\right] = -\left[A_i, \partial_0 A_i - \partial_i A_0 + \left[A_0, A_i\right]\right]$,

$$\Delta A_0 = -J_0 + \partial^i \left[A_0, A_i\right]$$
$$= 2\left[\partial^i A_0, A_i\right] + \left[A_i, \partial_0 A_i\right] + \left[A_i, \left[A_0, A_i\right]\right]$$

Setting $\alpha = i$ in (Y-M') we derive in the same fashion,

$$J_i = -\partial_t \mathbf{F}_{i0} + \partial_j \mathbf{F}_{ij}$$
$$= -\Box A_i - \partial_t \partial_i A_0 - \partial_t \left[A_i, A_0\right] + \partial_j \left[A_i, A_j\right]$$

where $\Box$ denotes the D'Alembertian operator $\Box = \partial^\alpha \partial_\alpha = -\partial_t^2 + \Delta$. Hence,

$$\Box A_i = -\partial_t \partial_i A_0 - \partial_t \left[A_i, A_0\right] + \partial_j \left[A_i, A_j\right] - J_i$$

or, since $J_i = \left[A_0, \partial_i A_0 - \partial_0 A_i + \left[A_i, A_0\right]\right] - \left[A_j, \partial_i A_j - \partial_j A_i + \left[A_i, A_j\right]\right]$,

$$\Box A_i + \partial_t \partial_i A_0 = -2\left[A_j, \partial_j A_i\right] + \left[A_j, \partial_i A_j\right] + \left[\partial_t A_0, A_i\right] + 2\left[A_0, \partial_t A_i\right]$$
$$- \left[A_0, \partial_i A_0\right] - \left[A_j, \left[A_j, A_i\right]\right] + \left[A_0, \left[A_0, A_i\right]\right].$$

We are therefore looking for solutions A_0, A of the following system of differential equations,

$$\Delta A_0 = 2\left[\partial^i A_0, A_i\right] + \left[A_i, \partial_0 A_i\right] + \left[A_i, \left[A_0, A_i\right]\right] \tag{3.11a}$$
$$\Box A_i + \partial_t \partial_i A_0 = -2\left[A_j, \partial_j A_i\right] + \left[A_j, \partial_i A_j\right] + \left[\partial_t A_0, A_i\right] + 2\left[A_0, \partial_t A_i\right]$$
$$- \left[A_0, \partial_i A_0\right] - \left[A_j, \left[A_j, A_i\right]\right] + \left[A_0, \left[A_0, A_i\right]\right] \tag{3.11b}$$
$$\nabla^i A_i = 0. \tag{3.11c}$$

Let $\mathcal{P}$ denote the projection operator on divergence free vectorfields, i.e.

$$\mathcal{P}B = (-\Delta)^{-1}\nabla \times \nabla \times B.$$

Using it we can rewrite

$$\Box A_i = V B_i + B_i + G_i \tag{3.12}$$

$$V B_i = \mathcal{P}\big[A_j, \partial_i A_j\big], \qquad B_i = -2\mathcal{P}\big[A_j, \partial_j A_i\big]$$

$$G_i = \mathcal{P}\Big(\big[\partial_t A_0, A_i\big] + 2\big[A_0, \partial_t A_i\big]$$

$$- \big[A_0, \partial_i A_0\big] + \big[A_0, [A_0, A_i]\big] - \big[A_j, [A_j, A_i]\big] \Big)$$

In view of the energy inequality (3.3) to control $\|\partial A(t, \cdot)\|_{L^2(\mathbf{R}^n)}$ we need to control $\int_0^t \|VB(t', \cdot)\|_{L^2(\mathbf{R}^n)} dt'$, $\int_0^t \|B(t', \cdot)\|_{L^2(\mathbf{R}^n)} dt'$ and $\int_0^t \|G(t', \cdot)\|_{L^2(\mathbf{R}^n)} dt'$. The last term can be estimated by standard methods. To estimate B we need the following:

Proposition 3.4. *Consider the following system of wave equations in $\mathbf{R}^{3+1}$,*

$$\Box A_i = F_i, \quad \Box \phi = f$$

subject to the initial conditions at $t = 0$,

$$A(0, \cdot) = a_{(0)}, \quad \partial_t A(0, \cdot) = a_{(1)}$$
$$\phi(0, \cdot) = \varphi_{(0)}, \quad , \partial_t \phi(0, \cdot) = \varphi_{(1)}$$

with $\operatorname{div} a_{(0)} = \operatorname{div} a_{(1)} = 0$. Assume also that the spatial divergence of F, $\operatorname{div} F$, is zero. Then,

$$\int\int_{[0,T]\times\mathbf{R}^3} \left| \big[A^j, \partial_j \phi\big] \right|^2 dx dt$$

$$\leq C \left(\|\nabla a_{(0)}\|_{L^2(\mathbf{R}^3)} + \|a_{(1)}\|_{L^2(\mathbf{R}^3)} + \int_0^T \|F(t, \cdot)\|_{L^2(\mathbf{R}^3)} \, dt \right)^2 \cdot \tag{3.13}$$

$$\cdot \left(\|\nabla \varphi_{(0)}\|_{L^2(\mathbf{R}^3)} + \|\varphi_{(1)}\|_{L^2(\mathbf{R}^3)} + \int_0^T \|f(t, \cdot)\|_{L^2(\mathbf{R}^3)} \, dt \right)^2$$

for any data $a_{(0)}, a_{(1)}, \varphi_{(0)}, \varphi_{(1)}$, and F, f which verify our assumptions and for which the right hand side of the above inequality is finite.

To prove the Proposition observe that, since F is divergence free we can write $F = \nabla \times \bar{F}$, $\operatorname{div} \bar{F} = 0$. Simmilarily the data $a_{(0)} = \nabla \times \bar{a}_{(0)}$, $a_{(1)} = \nabla \times \bar{a}_{(1)}$. Hence $A = \nabla \times \bar{A}$ where $\bar{A}$ verifies the equations,

$$\Box \bar{A} = \bar{F}$$

with the new initial data $\bar{a}_{(0)}, \bar{a}_{(1)}$. Now $[A_j, \partial_j \phi] = \epsilon_{iab} [\partial_a \bar{A}, \partial_i \phi]$. Observe that the last term is a linear combination of the null forms $Q_{ij}(\phi, \bar{A})$ defined in (3.8b).

The result is now an immediate application of Proposition 3.3 applied to the wave equations satisfied by $\bar{A}$ and ϕ.

To estimate VB we observe that $\in_{iab} \partial_a (VB)_b = \in_{iab} \left[\partial_a A_j, \partial_i A_j\right]$. Hence VB can also be expressed in terms of the null forms $Q_{ij}(A, A)$. We can thus estimate it using the following refinement of Proposition 3.3.

Proposition 3.5. *Under the same assumptions as in Proposition 3.3 , in $\mathbb{R}^{1+3}$ we have,*

$$\int_0^T \int_{\mathbb{R}^3} |(-\Delta)^{-\frac{1}{2}} Q(\phi, \psi)|^2 \, dxdt$$

$$\leq c \left(\|\nabla f_0\|_{L^2(\mathbb{R}^3)} + \|f_1\|_{L^2(\mathbb{R}^3)} + \int_0^T \|F(t, \cdot)\|_{L^2(\mathbb{R}^3)} \, dt \right)^2 \qquad (3.14)$$

$$\cdot \left(\|\nabla g_0\|_{L^2(\mathbb{R}^3)} + \|g_1\|_{L^2(\mathbb{R}^3)} + \int_0^T \|G(t, \cdot)\|_{L^2(\mathbb{R}^3)} \, dt \right)^2$$

The estimates (3.13) and (3.14) provide the main new analytic tool in proving Theorem 5 stated in section 2. Nevertheless there are some further serious difficulties to overcome. The main one concerns the use of the global Coulomb condition $\nabla^i A_i = 0$. It is well known that, for large data, the global Coulomb gauge leads to fundamental difficulties. In our paper [Kl-Ma2] we circumvent this problem by considering local Coulomb gauges adapted to finite past causal domains. This requires an appropriate boundary condition. The condition has to be chosen in such a way as to be able to localize the estimates (3.13–3.14). One also needs a method of passing from the estimates in the Coulomb gauge, which are local, to the temporal gauge in which the solutions can be globally extended.

4. Proof of the Null Estimates

In this section I shall sketch the proof of the Null estimates of the propositions 3.3, 3.5. We first observe that it suffices to consider the homogeneous case,

$$\Box\phi = 0, \qquad \Box\psi = 0 \qquad (4.1)$$

subject to the standard initial value problem,

$$\phi(0, x) = 0, \partial_t \phi(0, x) \quad = f(x)$$
$$\psi(0, x) = 0, \partial_t \psi(0, x) \quad = g(x).$$

The general case follows then by a simple application of the Duhamel's principle. Observe that, $\phi = \frac{1}{2i}(\phi_+ - \phi_-)$ where,

$$\phi_\pm(t, x) = (2\pi)^{-3} \int_{\mathbb{R}^3} e^{\pm it(|\xi| + ix \cdot \xi)} \frac{\hat{f}(\xi)}{|\xi|} \, d\xi \qquad (4.2)$$

with $\hat{f}(\xi)$ the Fourier transform of f. Let $\tilde{\phi}$ denote the space-time Fourier transform i.e., $\tilde{\phi}(\xi, \tau) = \int_{\mathbb{R}^{1+3}} \phi(x, t) e^{-i(x \cdot \xi + t\tau)} \, dx dt$. Then,

$$\tilde{\phi}_+(\tau, \xi) = \delta(\tau - |\xi|) \frac{\hat{f}(\xi)}{|\xi|}$$

$$\tilde{\phi}_-(\tau, \xi) = \delta(\tau + |\xi|) \frac{\hat{f}(\xi)}{|\xi|}$$

Proposition 3.3, resp. 3.5, is an immediate consequence of part (i), resp (iii), of the following:

Theorem 4.1. *Let ϕ, ψ be solutions of 4.1. Let $\dot{H}^s$ be the standard homogeneous Sobolev norms, $\|f\|_{\dot{H}^s} = \left(\int |\xi|^{2s} |\hat{f}(\xi)|^2 d\xi \right)^{\frac{1}{2}}$.*

i)   *For all dimensions $n \geq 2$ and all null forms 3.8a, 3.8b*

$$\|Q(\phi, \psi)\|_{L^2(\mathbb{R}^{1+n})} \leq c \|f\|_{L^2(\mathbb{R}^n)} \|g\|_{\dot{H}^{\frac{n-1}{2}}(\mathbb{R}^n)} \tag{4.3a}$$

ii) *For all dimensions $n \geq 3$ and all null forms 3.8a, 3.8b*

$$\left\| \frac{1}{(\tau^2 - |\xi|^2)^{\frac{1}{2}}} \tilde{Q}(\phi, \psi) \right\|_{L^2(\mathbb{R}^{1+n})} \leq c \|f\|_{L^2(\mathbb{R}^n)} \|g\|_{\dot{H}^{\frac{n-3}{2}}(\mathbb{R}^n)} \tag{4.3b}$$

iii) *For all dimensions $n \geq 3$ and all null forms Q_{ij},*

$$\left\| \frac{1}{|\xi|} \tilde{Q}(\phi, \psi) \right\|_{L^2(\mathbb{R}^{1+n})} \leq c \|f\|_{L^2(\mathbb{R}^n)} \|g\|_{\dot{H}^{\frac{n-3}{2}}(\mathbb{R}^n)} \tag{4.3c}$$

Before proving Theorem 4.1 we shall first review the proof of the classical Strichartz inequality in $L^4(\mathbb{R}^{1+3})$. The proof uses a method similar in spirit to that of [Ca-Sj].

Proposition 4.1. *Consider the homogeneous wave equation in $\mathbb{R}^{1+n}$, $\Box\phi = 0$, subject to the standard initial value problem, $\phi(0, \cdot) = 0$, $\partial_t \phi(0, \cdot) = f$. The classical, isotropic[36] Strichartz inequality in dimensions $n = 3$ reads as follows:*

$$\|\phi^2\|_{L^2(\mathbb{R}^{1+3})} \leq c \|f\|^2_{\dot{H}^{-\frac{1}{2}}}.$$

To prove the Proposition we first observe that it suffices to estimate ϕ_+. Indeed $\|\phi\|_{L^4} \leq \|\phi_+\|_{L^4} + \|\phi_-\|_{L^4}$. Now $\|\phi_+\|^2_{L^4} = \|(\phi_+)^2\|_{L^2} = c\|\tilde{\phi}_+ * \tilde{\phi}_+\|_{L^2}$

---

36) The general Strichartz, or Strichartz-Brenner, inequality refers to $L^{p,q}$ norms with different exponents for the space and time variables.

To prove the proposition it thus suffices to prove the following,

$$\left\|\tilde{\phi}_+ * \tilde{\phi}_+\right\|_{L^2} \le c\|\hat{f}\|_{\dot{H}^{-\frac{1}{2}}}^2 \tag{4.5}$$

Now,

$$\tilde{\phi}_+ * \tilde{\phi}_+(\tau, \xi) = \int \delta(\tau - \lambda - |\xi - \eta|) \frac{|\hat{f}(\xi - \eta)|}{|\xi - \eta|} \delta(\lambda - |\eta|) \frac{|\hat{f}(\eta)|}{|\eta|} d\lambda d\eta$$

$$= \int \delta(\tau - |\eta| - |\xi - \eta|) \frac{|\hat{f}'(\xi - \eta)|}{|\xi - \eta|^{\frac{1}{2}}} \frac{|\hat{f}'(\eta)|}{|\eta|^{\frac{1}{2}}} d\eta$$

where $\hat{f}' = |\xi|^{-\frac{1}{2}} \hat{f}(\xi) \in L^2(\mathbb{R}^n)$. It thus suffices to prove that, for all L^2 functions f,

$$\left\|\int \delta(\tau - |\eta| - |\xi - \eta|) \frac{|\hat{f}(\xi - \eta)|}{|\xi - \eta|^{\frac{1}{2}}} \frac{\hat{f}(\eta)|}{|\eta|^{\frac{1}{2}}} d\eta\right\|_{L^2(d\tau d\eta)} \le c\|f\|_{L^2}^2 \tag{4.6}$$

We can prove it by using Cauchy-Schwartz with respect to the measure $\delta(\tau - |\eta| - |\xi - \eta|)d\eta$,

$$\left|\int \delta(\tau - |\eta| - (\xi - |\eta|)) \frac{|\hat{f}(\xi - \eta)|}{|\xi - \eta|^{\frac{1}{2}}} \frac{|\hat{f}(\eta)|}{|\eta|^{\frac{1}{2}}} d\eta\right|^2$$

$$\le \int \delta(\tau - |\eta| - |\xi - \eta|) \frac{1}{|\xi - \eta|} \frac{1}{|\eta|} d\eta$$

$$\cdot \int \delta(\tau - |\eta| - |\xi - \eta|)|\hat{f}(\xi - \eta)|^2|\hat{f}(\eta)|^2 d\eta .$$

The first integral above is bounded uniformly in τ, ξ (see Lemma 4.1 below). Integrating the second one with respect to τ, ξ proves the estimate (4.6) and thus the Proposition.

Lemma 4.1. *The integral*

$$\int_{\mathbb{R}^3} \delta(\tau - |\xi - \eta| - |\eta|) \frac{1}{|\xi - \eta||\eta|} d\eta$$

is bounded uniformly for all values of $(\tau, \xi) \in \mathbb{R}^{1+3}$, $|\tau| \ge |\xi|$

The proof of Theorem 4.1 is similar. We start by writing,

$$\tilde{Q}(\tau, \xi) = \int b(\tau - \lambda, \xi - \eta; \lambda, \eta)\tilde{\phi}(\tau - \lambda, \xi - \eta)\tilde{\psi}(\lambda, \eta)d\lambda d\eta,$$

The symbol b of the corresponding null form is given by,

$$b = b_0(\tau, \xi; \lambda, \eta) = -\tau\lambda + \xi \cdot \eta \tag{4.7a}$$

in the case of Q_0,

$$b = b_{ij}(\tau, \xi; \lambda, \eta) = \xi_i \eta_j - \xi_j \eta_i \qquad (4.7b)$$

in the case of Q_{ij} and,

$$b = b_{0i}(\tau, \xi; \lambda, \eta) = \tau \eta_i + \lambda \eta_j \qquad (4.7c)$$

in the case of Q_{0i}.

The major difference between estimating null forms and the proof of Proposition 4.1 is that we now have to consider both $Q(\phi_+, \psi_+)$ and $Q(\phi_+, \psi_-)$.

We write:

$$\widetilde{Q(\phi_+, \psi_\pm)}(\tau, \xi)$$

$$= \int \int \frac{b(\tau - \lambda, \xi - \eta, \lambda, \eta) f(\xi - \eta) g(\eta)}{|\xi - \eta||\eta|} \delta(\tau - \lambda - |\xi - \eta|) \delta(\lambda \mp |\eta|) d\lambda d\eta \qquad (4.8)$$

$$= \int \frac{\sigma_\pm(\xi - \eta, \eta)}{|\xi - \eta||\eta|} \hat{f}(\xi - \eta) \hat{g}(\eta) \delta(\tau - |\xi - \eta| \mp |\eta|) d\lambda d\eta$$

where,

$$\sigma_\pm(\xi, \eta) = b(|\xi|, \xi, \pm|\eta|, \eta). \qquad (4.9)$$

Lemma 4.2 *Consider the null forms (3.8a, 3.8b) their full symbols (4.7a–4.7c) and reduced symbols defined by (4.9).*

(i) *In the case of the null form Q_0 we have,*

$$\sigma_\pm(\xi, \eta) = |\xi||\eta| \mp \xi \cdot \eta,$$

and

$$|\sigma_+(\xi, \eta)| \leq c(|\xi| + |\eta| - |\xi + \eta|)(|\xi| + |\eta|) \qquad (4.10a)$$

$$|\sigma_-(\xi, \eta)| \leq c(|\xi + \eta| - ||\xi| - |\eta||)(|\xi| + |\eta|) \qquad (4.10b)$$

(ii) *In the case of the null form Q_{ij} we have,*

$$\sigma_\pm(\xi, \eta) = \xi_i \eta_j - \xi_j \eta_i.$$

and,

$$|\sigma_\pm(\xi, \eta) \leq c|\xi|^{\frac{1}{2}}|\eta|^{\frac{1}{2}}|\xi + \eta|^{\frac{1}{2}} \min\left((|\xi| + |\eta| - |\xi + \eta| \, ; \, |\xi + \eta| - ||\xi| - |\eta||)\right) \qquad (4.10c)$$

(iii) *In the case of the null form Q_{0i} we have,*

$$\sigma_\pm(\xi, \eta) = |\xi|\eta_j \mp \xi_j|\eta|,$$

and

$$|\sigma_+(\xi, \eta)| \leq c|\xi|^{\frac{1}{2}}|\eta|^{\frac{1}{2}}(|\xi| + |\eta| - |\xi + \eta|)^{\frac{1}{2}}(|\xi| + |\eta|)^{\frac{1}{2}} \qquad (4.10d)$$

$$|\sigma_-(\xi, \eta)| \leq c|\xi|^{\frac{1}{2}}|\eta|^{\frac{1}{2}}(|\xi + \eta| - ||\xi| - |\eta||)^{\frac{1}{2}}(|\xi| + |\eta|)^{\frac{1}{2}} \qquad (4.10e)$$

The proof of the estimates (4.10a–4.10b) are an immediate consequence of the identities $2(|\xi||\eta| - \xi \cdot \eta) = (|\xi| + |\eta| - |\xi + \eta|)(|\xi| + |\eta| + |\xi + \eta|)$ and $2(|\xi||\eta| + \xi \cdot \eta) = (|\xi + \eta| - ||\xi| - |\eta||)(|\xi + \eta| + ||\xi| - |\eta||)$.

To prove the estimate 4.10c we shall make use of the identity, $|\xi \times \eta|^2 = \sum_{i<j}(\xi_i \eta_j - \xi_j \eta_i)^2 = |\xi|^2|\eta|^2 - (\xi \cdot \eta)^2$. Hence,

$$|\xi \times \eta| \le (|\xi||\eta| - \xi \cdot \eta)^{\frac{1}{2}}(|\xi||\eta| + \xi \cdot \eta)^{\frac{1}{2}}$$

Now, if $|\xi| + |\eta| - |\xi + \eta| \le |\xi + \eta| - ||\xi| - |\eta||$ then $|\xi|, |\eta| \le |\xi + \eta|$. Hence,

$$|\xi \times \eta| \le (|\xi||\eta| + \xi \cdot \eta)^{\frac{1}{2}}\frac{1}{\sqrt{2}}(|\xi| + |\eta| - |\xi + \eta|)^{\frac{1}{2}}(|\xi| + |\eta| + |\xi + \eta|)^{\frac{1}{2}}$$

$$\le (|\xi||\eta| + \xi \cdot \eta)^{\frac{1}{2}}|\xi + \eta|^{\frac{1}{2}}(|\xi| + |\eta| - |\xi + \eta|)^{\frac{1}{2}}$$

$$\le 2^{\frac{1}{2}}|\xi|^{\frac{1}{2}}|\eta|^{\frac{1}{2}}|\xi + \eta|^{\frac{1}{2}}(|\xi| + |\eta| - |\xi + \eta|)^{\frac{1}{2}}$$

On the other hand if $|\xi| + |\eta| - |\xi + \eta| \ge |\xi + \eta| - ||\xi| - |\eta||$ we have,

$$|\xi \times \eta| \le (|\xi||\eta| - \xi \cdot \eta)^{\frac{1}{2}}\frac{1}{\sqrt{2}}(|\xi + \eta| - ||\xi| - |\eta||)^{\frac{1}{2}}(|\xi + \eta| + ||\xi| - |\eta||)^{\frac{1}{2}}$$

$$\le 2^{\frac{1}{2}}|\xi|^{\frac{1}{2}}|\eta|^{\frac{1}{2}}|\xi + \eta|^{\frac{1}{2}}(|\xi + \eta| - ||\xi| - |\eta||)^{\frac{1}{2}}$$

For simplicity I shall only indicate how to prove the estimates (4.3b) and (4.3c) of Theorem 4.1 for the special case of the null form Q_{ij} and dimension $n = 3$. These are in fact the estimates needed in applications to the Yang-Mills equations. The other estimates are proved roughly in the same way. According to (4.8) and Lemma 4.2 we have:

$$\tilde{Q}(\tau, \xi) \le c||\tau| - |\xi||^{\frac{1}{2}}\int |\xi|^{\frac{1}{2}}|\eta|^{\frac{1}{2}}|\xi - \eta|^{\frac{1}{2}}\delta(\tau - |\xi - \eta| \mp |\eta|)\frac{\hat{f}(\xi - \eta)\hat{g}(\eta)}{|\xi - \eta||\eta|}d\eta$$

$$\le c||\tau| - |\xi||^{\frac{1}{2}}|\xi|^{\frac{1}{2}}\int \delta(\tau - |\xi - \eta| \mp |\eta|)\frac{\hat{f}(\xi - \eta)\hat{g}(\eta)}{|\xi - \eta|^{\frac{1}{2}}|\eta|^{\frac{1}{2}}}d\eta$$

Therefore,

$$\frac{\tilde{Q}(\tau, \xi)}{||\tau| - |\xi||^{\frac{1}{2}}|\xi|^{\frac{1}{2}}} \le c\widetilde{\Phi\Psi}(\tau, \xi) \tag{4.11}$$

where Φ, Ψ are both solutions of the homogeneous wave equation, $\Box\Phi = \Box\Psi = 0$ with initial data,

$$\hat{\Phi}(0, \xi) = 0, \quad \partial_t\hat{\Phi}(0, \xi) = |\xi|^{\frac{1}{2}}\hat{f}(\xi) \in \dot{H}^{-\frac{1}{2}}$$

$$\hat{\Psi}(0, \xi) = 0, \quad \partial_t\hat{\Psi}(0, \xi) = |\xi|^{\frac{1}{2}}\hat{g}(\xi) \in \dot{H}^{-\frac{1}{2}},$$

provided that $f, g \in L^2(\mathbb{R}^3)$. Therefore the estimate (4.3b), in this case, is now an immediate consequence of Proposition 4.1. The estimate (4.3c) follows from (4.3b) in the region $|\xi| \geq \frac{|\tau|}{2}$. On the other hand, the region $|\xi| \leq \frac{|\tau|}{2}$ is disjoint of the support of $Q(\phi_+, \psi_-)$, we therefore only need to estimate $Q(\phi_+, \psi_+)$ for $|\xi| \leq \frac{|\tau|}{2}$. This region can be handled by proceeding precisely as in the proof of Proposition 4.1.

5. The Proof of Theorem 4

As we have mentioned in Remark 2 of section 3 the L^2 space-time estimates of Proposition 3.3 don't combine well with the energy estimates (3.3–3.4). In this section we discuss a different approach, see [Kl-Ma4], based entirely on space-time estimates. The approach is similar to that used by Bourgain [B] and Kenig-Ponce-Vega [Ke-Po-Ve] for the Korteweg-de Vries equation.

Consider the equations (3.9),

$$\Box \phi^I + \sum_{J,K} \Gamma^I_{J,K}(\phi) B^I_{JK}(\partial \phi^J, \partial \phi^K) = 0 \tag{5.1}$$

with the B^I_{JK} any of the null forms (3.8a-3.8b),

$$Q_0(\phi, \psi) = \partial_\alpha \phi \cdot \partial^\alpha \psi = -\partial_t \phi \partial_t \psi + \sum_{i=1}^n \partial_i \phi \partial_i \psi \tag{5.2a}$$

$$Q_{\alpha\beta}(\phi, \psi) = \partial_\alpha \phi \partial_\beta \psi - \partial_\beta \phi \partial_\alpha \psi \quad 0 \leq \alpha < \beta \leq n. \tag{5.2b}$$

In the particular case when only the null form Q_0 is allowed to appear in (5.1), we may say that the corresponding equations are of "Wave Maps" type. Indeed the equations satisfied by Wave Maps ϕ defined from the Minkowski space-time $\mathbb{R}^{n+1}$ to a Riemannian manifold M, take precisely that form when expressed relative to a system of local coordinates in M for which the Γ's are the corresponding Christoffel symbols.

We consider the space-time norms,

$$N_{s,\delta}(\phi) = \left(\int w_+^{2s}(\tau, \xi) w_-^{2\delta}(\tau, \xi) \log^2(1 + w_-(\tau, \xi)) |\phi(\tau, \xi)|^2 d\tau d\xi \right)^{1/2} \tag{5.3}$$

where,

$$w_\pm(\tau, \xi) = 1 + \Big| |\tau| \pm |\xi| \Big|. \tag{5.3a}$$

Consider also the homogeneous weights,

$$\dot{w}_\pm(\tau, \xi) = \Big| |\tau| \pm |\xi| \Big|. \tag{5.3b}$$

The reason these norms are natural is that they capture the optimal gain of regularity of the solution of $\Box \phi = F$. Indeed, if we take χ to be a smooth cut-off function in time, we have

$$N_{s+1,\delta+1}(\chi\phi) \le C\left(\|\partial\phi(0,\cdot)\|_{H^s} + N_{s,\delta}(F)\right)$$

for all $\delta \ge -\frac{1}{2}$.

The norms are also compatible with the null forms (5.2a–5.2b). This can be seen already from the following,

Proposition 5.1.

(i)   *The symbol* $b_0(\tau,\xi;\lambda,\eta) = \tau\lambda - \xi\cdot\eta$ *of* Q_0 *verifies the estimate,*

$$|b_0| \le 2\left(\dot{w}_-(\tau,\xi)\dot{w}_+(\tau,\xi) + \dot{w}_-(\lambda,\eta)\dot{w}_+(\lambda,\eta) + \dot{w}_-(\tau+\lambda,\xi+\eta)\dot{w}_+(\tau+\lambda,\xi+\eta)\right)$$

(ii) *The symbol* $b_{ij}(\tau,\xi;\lambda\eta) = \xi_i\eta_j - \xi_j\eta_i$ *of* Q_{ij} *verifies the estimate,*

$$|b_{ij}| \le c|\xi|^{\frac{1}{2}}|\eta|^{\frac{1}{2}}|\xi+\eta|^{\frac{1}{2}}\left(\dot{w}_-(\tau,\xi)^{\frac{1}{2}} + \dot{w}_-(\lambda,\eta)^{\frac{1}{2}} + \dot{w}_-(\tau+\lambda,\xi+\eta)^{\frac{1}{2}}\right)$$

(iii) *The symbol* $b_{0i}(\tau,\xi;\lambda,\eta) = \tau\eta_i - \lambda\xi_i$ *of* Q_{0i} *verifies the estimate,*

$$|b_{oi}| \le |\eta|\dot{w}_-(\tau,\xi) + |\xi|\dot{w}_-(\lambda,\eta) + c|\xi|^{\frac{1}{2}}|\eta|^{\frac{1}{2}}(|\xi|+|\eta|)^{\frac{1}{2}}$$
$$\left(\dot{w}_-(\tau,\xi)^{\frac{1}{2}} + \dot{w}_-(\lambda,\eta)^{\frac{1}{2}} + \dot{w}_-(\tau+\lambda,\xi+\eta)^{\frac{1}{2}}\right)$$

for $\tau\cdot\lambda \ge 0$ *and,*

$$|b_{oi}| \le |\eta|\dot{w}_-(\tau,\xi) + |\xi|\dot{w}_-(\lambda,\eta) + c|\xi|^{\frac{1}{2}}|\eta|^{\frac{1}{2}}(|\xi+\eta|)^{\frac{1}{2}}$$
$$\left(\dot{w}_-(\tau,\xi)^{\frac{1}{2}} + \dot{w}_-(\lambda,\eta)^{\frac{1}{2}} + \dot{w}_-(\tau+\lambda,\xi+\eta)^{\frac{1}{2}}\right)$$

for $\tau\cdot\lambda \le 0$.

The proof of part i) follows directly from the identity,

$$2(\tau\lambda - \xi\cdot\eta) = (\tau+\lambda)^2 - |\xi+\eta|^2 - (\tau^2 - |\xi|^2) - (\lambda^2 - |\eta|^2).$$

The proof of part ii) follows from 4.10c as well as the following,

Lemma 5.2. *Let* $W = W(\tau,\xi;\lambda,\eta)$ *be the maximum of the weights* $\dot{w}_-(\tau,\xi)$, $\dot{w}_-(\lambda,\eta)$, $\dot{w}_-(\tau+\lambda,\xi+\eta)$. *Then*
(i)   *if* $\tau,\ \lambda \le 0$ *or* $\tau,\lambda \le 0$, *then* $\frac{1}{3}(|\xi|+|\eta|-|\xi+\eta|) \le W(\tau,\xi;\lambda,\eta)$
(ii) *if* $\tau \ge 0,\ \lambda \le 0$ *and* $\tau+\lambda \ge 0$ *or* $\tau \le 0,\ \lambda \ge 0$ *and* $\tau+\lambda \le 0$ *then* $\frac{1}{3}(|\eta|+|\xi+\eta|-|\xi|) \le W(\tau,\xi;\lambda,\eta)$
(iii) *if* $\tau \ge 0,\ \lambda \le 0$ *and* $\tau+\lambda \le 0$ *or* $\tau \le 0,\ \lambda \ge 0$ *and* $\tau+\lambda \ge 0$ $\frac{1}{3}(|\xi|+|\xi+\eta|-|\xi|) \le W(\tau,\xi:\lambda,\eta)$

Proof.

(i) Assume $\tau, \lambda \geq 0$. If either of $|\tau - |\xi||$ or $|\lambda - |\eta||$ is greater or equal, then $\frac{1}{3}(|\xi| + |\eta| - |\xi + \eta|)$ we are done. Thus, assume the opposite. Then,

$$|\xi| + |\eta| - |\xi - \eta| = (|\xi| - \tau) + (|\eta| - \lambda) + \tau + \lambda - |\xi + \eta|$$

$$< \frac{2}{3}(|\xi| + |\eta| - |\xi + \eta|) + |\tau + \lambda - |\xi + \eta||$$

Hence $|\tau + \lambda - |\xi + \eta||$ is greater than $\frac{1}{3}(|\xi| + |\eta| - |\xi + \eta|)$ and we are done. The case $\tau, \lambda \leq 0$ is proved in the same manner.

(ii) Assume $\tau \geq 0$, $\lambda \leq 0$ and $\tau + \lambda \geq 0$. If either $|\tau - |\xi||$ or $|\lambda + |\eta||$ is greater or equal $\frac{1}{3}(|\eta| + |\xi + \eta| - |\xi|)$ we are done. Assume the opposite. Then,

$$|\eta| + |\xi - \eta| - |\xi| = (|\eta| + \lambda) + (\tau - |\xi|) + |\xi + \eta| - \tau - \lambda$$

$$< \frac{2}{3}(|\eta| + |\xi + \eta| - |\eta|) + |\tau + \lambda - |\xi + \eta||$$

i.e. $|\tau + \lambda - |\xi + \eta|| > \frac{1}{3}(|\eta| + |\xi + \eta| - |\xi|)$, which proves the desired inequality. All other cases are proved in the same manner.

Corollary 1. *Let* $W = W(\tau, \xi; \lambda, \eta)$ *be defined as before. Then,*

(i) *if* $\tau \cdot \lambda \geq 0$,

$$\frac{1}{3}||\xi| - |\eta| - |\xi + \eta|| \leq W(\tau, \xi; \lambda, \eta)$$

(ii) *if* $\tau \cdot \lambda \leq 0$,

$$\frac{1}{3}(|\xi + \eta| - ||\xi| - |\eta||) \leq W(\tau, \xi; \lambda, \eta).$$

The main results of this section are contained in the following,

Theorem 5.1. *Consider the space-time norms (5.3.) and functions* ϕ, ψ *defined[37] in* $\mathbb{R}^{3+1}$.

(i) *For the null form* $Q = Q_0$ *and any[38]* $s > \frac{1}{2}$,

$$N_{s,-1/2}\Big(Q(\phi, \psi)\Big) \leq C N_{s+1,1/2}(\phi) N_{s+1,1/2}(\psi) \tag{5.4}$$

(ii) *For any space-time functions* $(\phi_i)_{i=1,\ldots,k+1}$ *and any polynomial* $P(\phi_1, \ldots, \phi_{k-1})$ *we have, for every* $s > \frac{1}{2}$,

$$N_{s,-1/2}\Big(P(\phi_1, \ldots, \phi_{k-1})Q_0(\phi_k, \phi_{k+1})\Big)$$

$$\leq C N_{s+1,1/2}(\phi_1) \cdot \ldots \cdot N_{s+1,1/2}(\phi_{k+1}).$$

(iii) *The estimate (i) fails for the null forms* $Q = Q_{\alpha\beta}(\phi, \psi)$, $0 \leq \alpha < \beta \leq 3$ *and all* s *in the range* $\frac{1}{2} < s < 1$.

---

37) For which the norms $N_{s+1,1/2}(\phi)$, $N_{s+1,1/2}(\psi)$ are well defined and finite.

38) For $\mathbb{R}^{1+2}$ the corresponding result holds for $s > 0$.

As application we use the estimates (i), (ii) of Theorem 5.1 to prove the following result concerning non-linear equations of the type (1–3.)

Theorem 5.2 *The initial value problem,*

$$\phi(0, x) = f_0(x), \quad \partial_t\phi(0, x) = f_1(x)$$

for non-linear wave equations of the form,

$$\Box\phi^I + \sum_{J,K} \Gamma^I_{J,K}(\phi)Q_0(\phi^J, \phi^K) = 0,$$

where $\Gamma^I_{J,K}(\phi)$ are real analytic in $\phi = (\phi^1, \dots \phi^N)$, is well posed for $f_0 \in H^{\frac{3}{2}+\epsilon}$, $f_1 \in H^{\frac{1}{2}+\epsilon}$.

The result of Theorem 5.2 is optimal. Indeed this can easily be seen by considering the scalar equation $\Box\phi = Q_0(\phi, \phi)$ which can be solved exactly, see [Kl-Ma1].

Proof of Theorem 5.1. The main ideas in the proof of Theorem 5.1 are the following. Using duality, the estimate (i) is equivalent to

$$\int\int w_+^s w_-^{-\frac{1}{2}} \log(1 + w_-)\tilde{Q}H(\tau, \xi)d\tau d\xi \le CN_{s,1/2}(\phi)N_{s,1/2}(\psi)\|H\|_{L^2(\mathbb{R}^{1+3})}.$$

On the other hand, introducing,

$$F = w_+^{s+1}w_-^{\frac{1}{2}} \log(1 + w_-)\tilde{\phi}$$

$$G = w_+^{s+1}w_-^{\frac{1}{2}} \log(1 + w_-)\tilde{\psi}$$

and writing,

$$\tilde{Q}(\tau, \xi) = \int b(\tau - \lambda, \xi - \eta; \lambda, \eta)\tilde{\phi}(\tau - \lambda, \xi - \eta)\tilde{\psi}(\lambda, \eta)d\lambda d\eta,$$

where $b(\tau, \xi; \lambda, \eta)$ is the symbol corresponding to the null forms, we can reexpress the inequality (5.4) in the form,

$$I = \int\int \frac{b(\tau - \lambda, \xi - \eta; \lambda, \eta)}{w_-^{\frac{1}{2}}(\tau, \xi)w_-^{\frac{1}{2}}(\tau - \lambda, \xi - \eta)w_-^{\frac{1}{2}}(\lambda, \eta)} \frac{w_+^s(\tau, \xi)}{w_+^{s+1}(\tau - \lambda, \xi - \eta)w_+^{s+1}(\lambda, \eta)}$$

$$\frac{\log(1 + w_-(\tau, \xi))}{\log(1 + w_-(\tau - \lambda, \xi - \eta))\log(1 + w_-(\lambda, \eta))}F(\tau - \lambda, \xi - \eta)G(\lambda, \eta)$$

$$H(\tau, \xi)d\tau d\xi d\lambda d\eta \le C\|F\|\|G\|\|H\|,$$

$$(5.5)$$

where $\|\ \|$ denotes the L^2 norm in $\mathbb{R}^{1+3}$. The symbol b of the null form is given by,

$$b = b_0(\tau, \xi; \lambda, \eta) = -\tau\lambda + \xi \cdot \eta$$

in the case of Q_0.

After a change of variables,

$$I = \int \int \frac{b(\tau,\xi;\lambda,\eta)}{w_-^{\frac{1}{2}}(\tau+\lambda,\xi+\eta) w_-^{\frac{1}{2}}(\tau,\xi) w_-^{\frac{1}{2}}(\lambda,\eta)} \frac{w_+^s(\tau+\lambda,\xi+\eta)}{w_+^{s+1}(\tau,\xi) w_+^{s+1}(\lambda,\eta)}$$
$$\frac{\log(2+w_-(\tau+\lambda,\xi+\eta))}{\log(1+w_-(\tau,\xi))\log(1+w_-(\lambda,\eta))} F(\tau,\xi) G(\lambda,\eta) H(\tau+\lambda,\xi+\eta) d\tau d\xi d\lambda d\eta$$

Next, we make use of part (i) of Proposition 5.1,

$$b_0(\tau,\xi;\lambda,\eta)^{\frac{1}{2}} \le \left(w_-(\tau+\lambda,\xi+\eta)^{\frac{1}{2}} w_+(\tau+\lambda,\xi+\eta)^{\frac{1}{2}} \right.$$
$$\left. + w_-(\tau,\xi)^{\frac{1}{2}} w_+(\tau,\xi)^{\frac{1}{2}} + w_-(\lambda,\eta)^{\frac{1}{2}} w_+(\lambda,\eta)^{\frac{1}{2}} \right)$$

Therefore,

$$I \le I_1 + I_2 + I_3$$

where,

$$I_1 = \int \int \frac{|b(\tau,\xi;\lambda,\eta)|^{\frac{1}{2}}}{w_-^{\frac{1}{2}}(\tau,\xi) w_-^{\frac{1}{2}}(\lambda,\eta)} \frac{w_+^{s+\frac{1}{2}}(\tau+\lambda,\xi+\eta)}{w_+^{s+1}(\tau,\xi) w_+^{s+1}(\lambda,\eta)}$$
$$\frac{\log(2+w_-(\tau+\lambda,\xi+\eta))}{\log(1+w_-(\tau,\xi))\log(1+w_-(\lambda,\eta))} F(\tau,\xi) G(\lambda,\eta) H(\tau+\lambda,\xi+\eta) d\tau d\xi d\lambda d\eta$$

$$I_2 = \int \int \frac{|b(\tau,\xi;\lambda,\eta)|^{\frac{1}{2}}}{w_-^{\frac{1}{2}}(\tau+\lambda,\xi+\eta) w_-^{\frac{1}{2}}(\lambda,\eta)} \frac{w_+^s(\tau+\lambda,\xi+\eta)}{w_+^{s+\frac{1}{2}}(\tau,\xi) w_+^{s+1}(\lambda,\eta)}$$
$$\frac{\log(2+w_-(\tau+\lambda,\xi+\eta))}{\log(1+w_-(\tau,\xi))\log(1+w_-(\lambda,\eta))} F(\tau,\xi) G(\lambda,\eta) H(\tau+\lambda,\xi+\eta) d\tau d\xi d\lambda d\eta$$

$$I_3 = \int \int \frac{|b(\tau,\xi;\lambda,\eta)|^{\frac{1}{2}}}{w_-^{\frac{1}{2}}(\tau+\lambda,\xi+\eta) w_-^{\frac{1}{2}}(\tau,\xi)} \frac{w_+^s(\tau+\lambda,\xi+\eta)}{w_+^{s+1}(\tau,\xi) w_+^{s+\frac{1}{2}}(\lambda,\eta)}$$
$$\frac{\log(2+w_-(\tau+\lambda,\xi+\eta))}{\log(1+w_-(\tau,\xi))\log(1+w_-(\lambda,\eta))} F(\tau,\xi) G(\lambda,\eta) H(\tau+\lambda,\xi+\eta) d\tau d\xi d\lambda d\eta$$

To estimate the first term I_1 we use the foliation

$$u = |\tau| - |\xi| = \pm\tau - |\xi|$$
$$v = |\lambda| - |\eta| = \pm\lambda - |\eta| \tag{5.6}$$

thus $\tau = \pm(|\xi|+u)$, $\lambda = \pm(|\eta|+v)$. We can thus reduce the integral to one of the type which we have analyzed in [Kl-Ma1]. The integrals I_2, I_3 are treated in the same manner after a simple change of variables. More precisely our estimates depend on the following Lemmas.

Lemma 5.3. *Consider the integral,*

$$\mathcal{J}_s(f,g,H) = \int\int \frac{\left|\pm 1 - \hat{\xi}\cdot\hat{\eta}\right|^{\frac{1}{2}}}{(1+|\eta|)^{s+\frac{1}{2}}} f(\xi)g(\eta)H(|\xi|\pm|\eta|,\xi+\eta)d\xi\,d\eta$$

where $\hat{\xi} = \frac{\xi}{|\xi|}, \hat{\eta} = \frac{\eta}{|\eta|}$ *and* $f,g \in L^2(\mathbb{R}^3)$, $H \in L^2(\mathbb{R}^{1+3})$. *Then, for any* $s \geq \frac{1}{2}$,

$$\mathcal{J}_s(f,g,H) \leq C\|f\|_{L^2(\mathbb{R}^3)}\|g\|_{L^2(\mathbb{R}^3)}\|H\|_{L^2(\mathbb{R}^{1+3})}.$$

and

Lemma 5.4. *Consider the integral,*

$$\mathcal{J}_{a,b}(f,g,H)$$

$$= \int\int \frac{\left||\xi||\eta| \mp \xi\cdot\eta\right|^{\frac{1}{2}}}{(1+||\xi|\pm|\eta||+|\xi+\eta|)^a(1+|\eta|)^b} f(\xi)g(\eta)H(|\xi|\pm|\eta|,\xi+\eta)d\xi\,d\eta$$

where $f,g \in L^2(\mathbb{R}^3)$, $H \in L^2(\mathbb{R}^{1+3})$. *Then, for any* $a,b \geq 0$, $a+b \geq 2$, *and* $\frac{1}{2} \leq a \leq 1$

$$\mathcal{J}_{a,b}(f,g,H) \leq C\|f\|_{L^2(\mathbb{R}^3)}\|g\|_{L^2(\mathbb{R}^3)}\|H\|_{L^2(\mathbb{R}^{1+3})}.$$

Both Lemmas are used to treat the "hyperbolic" regions $|u|, |v|$ small. On the other hand, in the "elliptic" regions where $|u|$ or $|v|$ are large we use

Lemma 5.5. *Consider the integral,*

$$J_{a,b}(f,g,h) = \int\int \frac{1}{(1+|\xi|)^a(1+|\eta|)^b} f(\xi)g(\eta)h(\xi+\eta)d\xi\,d\eta$$

for any functions $f,g,h \in L^2(\mathbb{R}^3)$ *and numbers* $a,b \geq 0$. *Then, if* $a+b > \frac{3}{2}$,

$$J_{a,b}(f,g,h) \leq C\|f\|_{L^2(\mathbb{R}^3)}\|g\|_{L^2(\mathbb{R}^3)}\|h\|_{L^2(\mathbb{R}^3)}$$

Part (ii) of Theorem 5.1 is proved by similar means. On the other hand part (iii) is somewhat surprising in view of the fact that Proposition 5.1 is valid for all null forms. Our counterexample depends on one hand on the lack of an identity such as the one used in Proposition 5.1 (i), and on the other hand on a subtle interaction between the '+' and '−' regions of our foliation (5.6).

6. Conclusions

The methods presented in sections 3–5 have allowed us to make significant progress towards the resolution of the Conjectures 1 and 2 for semilinear Field Theories such as Yang-Mills and Wave Maps in Minkowski space-time. I will end my Lectures by pointing out what are the main unresolved issues.

(i)   As mentioned above the estimate 5.4 is false for all other null forms $\neq Q_0$. Nevertheless the norms 5.3 seem hard to avoid. One has some freedom to play with the hyperbolic exponent δ; instead of estimating $N_{s,-\frac{1}{2}}(Q(\phi,\psi))$ as in 5.4 one can try instead a norm $N_{s,\delta}(Q(\phi,\psi))$ with $-\frac{1}{2} < \delta < 0$. This idea was recently used by Yi Zhou [Z]. Modifying appropriately our proof of Theorem 5.1 he was able to improve the result of Theorem 3.3 by gaining an additional $\frac{1}{4}$ in both $\mathbb{R}^{2+1}$ and $\mathbb{R}^{3+1}$. While his result in $\mathbb{R}^{2+1}$ is sharp the result in $\mathbb{R}^{3+1}$ is still off by $\frac{1}{4}$. Do we have to give up altogether on the norms 5.3? Recently Machedon and I [Kl-Ma 6] were able to show that, in a slightly different situation when the estimates corresponding to 5.3 fail, the nonlinear problem admits nevertheless solutions in the spaces $N_{s,-\frac{1}{2}}$ for exponents s arbitrarily close to the optimal one. The method uses to a higher degree the nonlinear character of the equations. We hope that similar techniques can be applied for the Yang-Mills equations.

(ii)   The methods discussed above will, I believe, prove Conjecture 2 for all exponents $s > \bar{s}$. To obtain results for the optimal exponent $s = \bar{s}$ we need, on one hand, to suitably modify the norms 5.3. and, on the other hand, to use more of the geometric properties of the equations at hand.

(iii)   The case of quasilinear Field Theories, such as the Einstein field equations, is completely open. So far the best "well posedness" results available are those derived by classical energy estimates.

Bibliography

[B]   J. Bourgain, "Fourier Transform restriction phenomena for certain lattice subsets and applications to nonlinear evolution equations", *Geometric and Functional Anal.* **3** (1993) 107–156, 209–262.

[Bour] J.P. Bourguignon, "Stabilité par déformation non-linéaire de la metrique de Minkowski" *Séminaire N. Bourbaki* **1990–1991** June 1991.

[Br1]   C. Bruhat, "Théorème d'existence pour certain systèmes d'èquations aux dérivées partielles nonlinéaires", *Acta Matematica* **88** (1952) 141 225.

[Br2]   C. Bruhat, "Un théorème d'instabilité pour certains équations hyperboliques non-linéaires" *C.R. Acad. Sci. Paris* **276A** (1973) pp. 281.

[Ca-Sj]   L. Carlesson-P. Sjolin "Oscillatory integrals and the multiplier problem for the disk" *St. Math.* **44** (1972) 287–299.

[Ch1]   D. Christodoulou "Global solutions of nonlinear hyperbolic equations for small data" *Comm. Pure Appl. Math.* **39** (1986) 267–282.

[Ch2] D. Christodoulou, "A Mathematical Theory of Gravitational Collapse" *Comm. Math. Ph.* **109** (1987) p. 613.

[Ch3] D. Christodoulou, "The Formation of Black Holes and Singularities in Spherically Symmetric Gravitational Collapse" *Comm. Pure Appl. Math.* **44** (1991) 339–373.

[Ch-Kl1] D. Christodoulou-S. Klainerman, "Asymptotic Properties of Linear Field Equations in Minkowski Space-Time" *Comm.Pure Appl.Math.* **43** (1990) 137–199.

[Ch-Kl2] D. Christodoulou-S. Klainerman *The nonlinear stability of the Minkowski space-time.* Princeton Univ. Press 1993.

[Ch-Za] D. Christodoulou-A. Shadi Tahvildar-Zadeh, "Regularity of Spherically Symmetric Harmonic Maps of the $2 + 1$ dim. Minkowski Space." *Duke Math. J.* **71** (1993) 31–69.

[E-M] D. Eardley-V. Moncrief "The Global Existence of Yang-Mills-Higgs fields in $\mathbf{M}^{3+1}$" *C.M.P.* **83** (1982) 171–212.

[Ge] R. Geroch, *Asymptotic structure of space-time* P. Esposito and L. Witten (eds.), Plenum, New York, 1976.

[G-V] J. Ginibre-G. Velo "The global Cauchy problem for the nonlinear Klein-Gordon Equations" *Math. Z.* **180**(1985), 487–505.

[Gr1] M. Grillakis "Regularity and Asympt. Behavior of the Wave Eq. with a critical nonlinearity" *Ann. of Math.* **132** (1990) , 485–509.

[Gr2] M. Grillakis "Classical solutions for the Equivariant Wave Maps in 1+2 dimensions" preprint.

[Gu] C. Gu, "On the Cauchy problem for harmonic maps defined on two-dimensional Minkowski space", *Comm. Pure Apl. Math.* **33** (1980) 727–737.

[Jö] K. Jörgens, *Math. Z.*, "Das Angfangswertproblem im Grossen für eine Klasse nichtlinearer Wellengleichungen **77**, 295–307 (1961).

[Ka1] L. V. Kapitansky "Some generalizations of the Strichartz-Brenner inequality" *Leningrad Math. Jour. 1* (1990), no. 3, , 693–726.

[Ka2] L. V. Kapitansky "Global and unique weak solutions of nonlinear wave equations" Math. Res. Letters **1**, 211–223 (1994).

[Kl1] S. Klainerman "Long-time behavior of solutions to nonlinear wave equations" *Proc. of Int. Congress for Mathematicians*, Warsaw 1982.

[Kl2] S. Klainerman "The null condition and global existence to nonlinear wave equations" *Lectures on Appl. Math.*, **23** (1986) 293–326.

[Kl-Ma1] S. Klainerman-M. Machedon "Space-Time estimates for null forms and the local existence theorem", *Comm. Pure Appl. Math.* **46** (1993), 1221–1268

[Kl-Ma2] S. Klainerman-M. Machedon "Finite energy solutions of the Yang-Mills equations in $\mathbf{R}^{3+1}$." *Annals of Math.* **142** (1995), 39–119

[Kl-Ma3] S. Klainerman-M. Machedon "On the regularity properties of the Wave Equation" *Physics on Manifolds*, 1994 Kluwer Academic Publishhers edited by M. Flato, R. Kerner, A. Lichnerowicz.

[Kl-Ma4] S. Klainerman-M. Machedon "Smoothing Estimates for Null Forms and Applications" *Duke Math J.* **81** (1995) 99–133.

[Kl-Ma5] S. Klainerman-M. Machedon "Remark on an extension of Strichartz type inequalities", preprint.

[Kl-Ma6] S. Klainerman-M. Machedon "Hyperbolic Sobolev Norms and optimal local existence for a class of non-linear wave equations", in preparation.

[Ko]   M. Kovalyov "Long time behavior of a system of non-linear wave equations" Comm. P.D.E. 12 (1987) 471–501.

[Ke-Po-Ve]   C. Kenig, G. Ponce, L. Vega, "The Cauchy problem for the Korteweg-De Vries equation in Sobolev spaces of negative indices" *Duke Math. Journal* **71** No 1. pp. 1–21 (1994).

[L]   H. Lindblad "Counterexamples to local existence for semilinear wave equations" To appear in Amer. J. Math. (1996).

[L-S]   H. Lindblad-C.D. Sogge "On Existence and Scattering with minimal regularity for semilinear wave equations" *J. Funct. Anal.* **130** (1995), 357–426.

[P]   H. Pecher, "Nonlinear Small Data Scattering for the Wave and Klein-Gordon Equation", *Math. Z.*, **185** (1984) 261–270.

[Po-Si]   G. Ponce-T. Sideris, "Local Regularity of Nonlinear Wave Equations in Three Space Dimensions", *Comm. P.D.E* **18** (1993), 169–177.

[S-Y]   R. Schoen and S.-T. Yau, "On the proof of the positive math conjecture in general relativity" *Comm. Math. Physics* **65**(1) (1979) 45–76.

[Sh]   J. Shatah "Weak solutions and development of singularities in the $SU(2)-\sigma$ model" Comm. Pure Appl. Math. **41** (1988) 459–469.

[Sh-Za1]   J. Shatah-A. Shadi Tahvildar-Zadeh "Non-uniqueness and development of singularities for harmonic maps of the Minkowski space"

[Sh-Za2]   J. Shatah-A. Shadi Tahvildar-Zadeh "Regularity of harmonic maps from the Minkowski space into rotationally symmetric manifolds" *Comm. Pure Appl. Math.* **45** (1992) 947–971.

[Sh-Stru1]   J. Shatah-M. Struwe "Regularity results for nonlinear wave equations" *Annals of Math.* (2) **138** (1993), 503–518.

[Sh-Stru2]   J. Shatah-M. Struwe "Well Posedness in the energy space for Semilinear Wave Equations with critical growth" *Int. Math. Res. Notices* **7** (1994) 303–309.

[Si]   T. Sideris "Global existence of harmonic maps in Minkowski space" Comm. Pure Appl. Math. **42** (1989) 1–13.

[Str]   W. Strauss "Nonlinear Wave Equations" *Regional Conference series in mathematics*, nr. 73.

[Stru]   M. Struwe "Semilinear wave equations" *Bull. Amer. Math. Soc.* (N.S.) **26**(1992), 53–85.

[Z]   Yi Zhou "Local Existence with optimal regularity for Nonlinear wave equations" preprint.

Progress in Nonlinear Differential Equations
and Their Applications, Vol. 29
© 1997 Birkhäuser Verlag Basel/Switzerland

Static and Moving Vortices in Ginzburg-Landau Theories

FANG-HUA LIN

Courant Institute, New York University

ABSTRACT.
Lecture 1:
After a brief introduction to the history of the theory of superconductivity,
I shall describe the main mathematical result due to Bethuel-Brézis-Hélein
and recent improvements and simplifications. In particular the asymptotic
behavior of distributions of vortices for static solutions or approximate solutions.
Lecture 2:
The aim of this lecture is to demonstrate the important role played by the
so-called renormalized energy in the study of vortices. I shall explain the
connection between the critical points renormalized energy and solutions
to Ginzburg-Landau equations. Using this we can show certain histeresis
phenomena for the phase transition near the lower critical magnetic field.
Lecture 3:
Here I shall start with the Gor'kov-Eliashberg equation for the evolution
of Ginzburg-Landau. The global existence of classical solutions and their
asymptotic behavior follow from existencing results. We are particularly interested in the dynamical properties of vortices. Several well-known specu-
lations should be verified.

It is obvious that three lectures can hardly cover any topic with relative
completeness in this fascinating and fast growing subject. Nonetheless I shall
try to give a rough state of the art survey, and mention various unsolved,
certainly more challenging problems.

Introduction

This write-up covers the author's three Lectures given in March of 1995 at the
University of Tennessee, Knoxville for the Barrett Lectures series. The purpose
of these lectures was to give a brief description of some rigorous mathematical
work concerning the static and moving vortices of solutions to Ginzburg-Landau
equations arising in the theory of superconductivity. This short presentation was
relatively complete and comprehensive at that time. Since then, the literature on
the subject related to these discussions has almost doubled. I have no intention
here to give a complete survey. Indeed, it would not be very wise to do so on
a subject that is still fast growing. Instead, I shall simply write up, with some

necessary details, what was covered in these three lectures. I should, however, take this opportunity to make a few comments on some closely related research which has been done since then in the area covered by each lecture.

It was an honor for me to give a Barrett Lecture series. I would like to thank the Department of Mathematics, University of Tennessee for giving me the opportunity. In particular, I would like to thank the organizers, Professor G. Baker and Professor A. Freire for their assistance and warm hospitality.

Lecture 1.

1. Background and Models

The response of a superconducting material to an externally imposed magnetic field is most conveniently described by the diagram below, which shows the minimum energy state of the superconductor as a function of H_0, the applied magnetic field, and the dimensionless material parameter κ (known as the Ginzburg-Landau parameter). The parameter κ determines the type of superconducting material. For $\kappa < \frac{1}{\sqrt{2}}$, type-I superconductors, there is a critical magnetic field H_C below which the material will be the superconducting state, but above which it will be in the normal state. For $\kappa > \frac{1}{\sqrt{2}}$, type-II superconductors, there is a third state known as the mixed (or vortex) state. The mixed state consists of many normal filaments embedded in a superconducting matrix. Each of these filaments carries with it a quantized amount of magnetic flux, and is circled by a vortex of superconducting current. Thus these filaments are often known as vortex lines. One of the most challenging problems to mathematicians working on the superconductivity models is the understanding of vortex phenomena [GO] in type-II superconductors, which include the recently discovered high-temperature superconductors.

The transition from the normal state to the mixed state takes place by a bifurcation as the magnetic field is lowered through some critical value H_{C_2}. The critical field H_{C_1}, on the other hand, is calculated so that at this field the energy of the wholly superconducting solution becomes equal to the energy of the single vortex filament solution for an infinite superconductor.

The vortex structures have been studied extensively on the mezoscale using the well-known Ginzburg-Landau models of superconductivity [DMG, COH]. The existence of vortex-like solutions for the full nonlinear Ginzburg-Landau equations have been substantiated through many arguments ranging from asymptotic analysis to numerical simulations. However, it remains to be justified by rigorous mathematical analysis. Much progress has been made in recent years on establishing a mathematical framework for a rigorous description of both the static and dynamic properties of the vortex solutions; in particular, as the coherence length tends to zero (κ goes to infinity), various results have been obtained. From a technological point of view, this is of interest since recently discovered high critical temperature superconductors are known to have large values of κ, say κ in excess of 50.

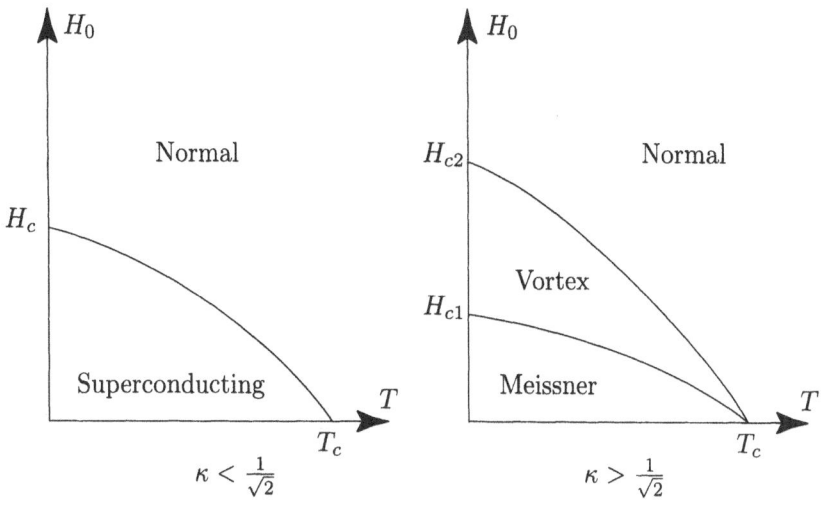

Figure 1: The diagram of various states for type-II superconductors

The vortex structures can be set in motion by a variety of mechanisms, including thermal fluctuations and applied voltages and currents. Unfortunately, such vortex motion in an applied magnetic field induces an effective resistance in the material, and thus a loss of superconductivity. Therefore, it is a crucial and rather difficult issue to understand the dynamics of these vortex lines. At the same time, one is interested in studying mechanisms that can pin the vortices at a fixed location i.e., prevent their motion. Various mechanisms have been advanced by physicists, engineers and mathematical scientists. For example, normal (non-superconducting) impurities in an otherwise superconducting material sample are believed to provide sites at which vortices are pinned. Likewise, regions of the sample that are thin relative to other regions are also believed to provide pinning sites. These mechanisms have been introduced into the general Ginzburg-Landau framework [DGP] to derive various variants of the original Ginzburg-Landau models of superconductivity. Numerical simulation clearly suggests the pinning effect.

Before addressing the above problems, we introduce the notations and the models that will be used in the lectures. The starting point of our study is the phenomenological model due to Ginzburg and Landau for superconductivity in isotropic, homogeneous material samples. Let Ω be a smooth bounded domain in R^3, occupied by the superconducting material. By ignoring the effect of the region exterior to the sample, the steady state model can be stated as a minimization problem of the free energy functional:

$$\mathcal{G}(\psi, \mathbf{A}) = \int_\Omega \left\{ f_n + a|\psi|^2 + \frac{b}{2}|\psi|^4 + \frac{1}{2m_s} \left| \left(i\hbar\nabla + \frac{e_s}{c}\mathbf{A} \right) \psi \right|^2 \right.$$
$$\left. + \frac{\mu_s}{8\pi} \mathbf{h} \cdot (\mathbf{h} - 2\mathbf{H}) \right\} d\Omega, \tag{1.1}$$

where f_n is the free energy density of the non-superconducting state in the absence of a magnetic field, ψ is the (complex-valued) superconducting order parameter, $\mathbf{A}$ is the magnetic vector potential, $\mathbf{h} = (1/\mu_s)\,\mathbf{curl\,A}$ is the magnetic field, $\mathbf{H}$ is the applied magnetic field, a and b are constants whose values depend on the temperature and such that $b > 0$, m_s is the mass and e_s is the charge of the superconducting charge carriers which is twice the electronic charge, e, c is the speed of light, μ_s is the permeability, $2\pi\hbar$ is Planck's constant. It can be rewritten in nondimensionalized form:

$$\mathcal{G}(\psi, \mathbf{A}) = \int_\Omega \left(\frac{1}{2}\left(1 - |\psi|^2\right)^2 + \left|\left(\frac{i}{\kappa}\nabla + \mathbf{A}\right)\psi\right|^2 + |\mathbf{curl A} - \mathbf{h}|^2 \right) dx \qquad (1.2)$$

where κ is the so-called Ginzburg-Landau parameter.

The functional $\mathcal{G}(\psi, \mathbf{A})$ has an interesting gauge invariance property and the minimization of $\mathcal{G}$ in appropriate functional spaces gives the following system of nonlinear differential equations that are referred to as the Ginzburg-Landau equations.

$$\left(\frac{i}{\kappa}\nabla + \mathbf{A}\right)^2 \psi - \psi + |\psi|^2\psi = 0 \quad \text{in} \quad \Omega \qquad (1.3)$$

$$\mathbf{curl\,curl\,A} = \frac{i}{2\kappa}\left(\psi\nabla\psi^* - \psi^*\nabla\psi\right) - |\psi|^2\mathbf{A} \quad \text{in} \quad \Omega \qquad (1.4)$$

along with natural boundary conditions

$$\mathbf{curl\,A} \wedge \mathbf{n} = \mathbf{h} \wedge \mathbf{n} \quad \text{on} \quad \Gamma = \partial\Omega \qquad (1.5)$$

and

$$\left(\frac{i}{\kappa}\nabla\psi + \mathbf{A}\psi\right) \cdot \mathbf{n} = 0 \quad \text{on} \quad \Gamma = \partial\Omega \qquad (1.6)$$

where $\mathbf{n}$ is the exterior normal to the boundary Γ.

The time-dependent Ginzburg-Landau model is often described by the Gor'kov-Eliashberg evolution equation:

$$\left.\begin{array}{l}
\eta\dfrac{\partial\psi}{\partial t} \;+\; i\eta\kappa\,\Phi\psi + \left(\dfrac{i}{\kappa}\nabla + \mathbf{A}\right)^2 \psi - \psi + |\psi|^2\psi = 0, \\[4mm]
\dfrac{\partial\mathbf{A}}{\partial t} \;+\; \nabla\Phi + \mathbf{curl\,curl\,A} = -\dfrac{i}{2\kappa}\left(\psi^*\nabla\psi - \psi\nabla\psi^*\right) - \mathbf{A}|\psi|^2,
\end{array}\right\} \qquad (1.7)$$

Here Φ denotes the (real) scalar electric potential, η is a relaxation parameter and ψ^* denotes the complex conjugate of ψ.

The system is supplemented by the initial and boundary conditions:

$$\psi(x, 0) = \psi_0(x), \quad \mathbf{A}(x, 0) = \mathbf{A}(x), \quad x \in \Omega; \qquad (1.8)$$

$$\left(\frac{i}{\kappa}\nabla + \mathbf{A}\right)\psi \cdot \mathbf{n} = 0,$$

$$\operatorname{curl}\mathbf{A}\wedge\mathbf{n} = \mathbf{h}\wedge\mathbf{n},$$

$$\left(\frac{\partial\mathbf{A}}{\partial t} + \nabla\Phi\right)\cdot\mathbf{n} = \vec{E}\cdot\mathbf{n} = 0, \quad \text{on} \quad \partial\Omega.$$

$$(1.9)$$

Note that (1.7) is Gauge-invariant, in the sense that if $(\psi, \mathbf{A}, \Phi)$ is a solution, then so it $(\psi_\chi, \mathbf{A}\chi, \Phi_\chi)$, where

$$\psi_\chi = \psi\, e^{i\,K\,\chi}, \quad \mathbf{A}\chi = \mathbf{A} + \nabla\chi, \quad \Phi_\chi = \Phi - \frac{\partial\chi}{\partial t}. \qquad (\text{cf. [D]})$$

For type-II superconductors, $\kappa > 1/\sqrt{2}$. The minimizers of $\mathcal{G}$ are believed to exhibit vortex structures. Numerical experiments have shown that for large values of κ and moderate field strengths, the number of vortices could be exceedingly large even for small sample size in actual physical scale. Thus, resolving the vortex phenomenon using the full Ginzburg-Landau equations remains computationally intensive.

Various simplications have been made to reduce the complexity. For thin films of superconducting material, a two-dimensional model has been developed [DG, CDG] that can account for thickness variations through an averaging process. The model is given by the following minimization problem:

$$\mathcal{G}_\epsilon^a(\psi) = \int_\Omega a(\mathbf{x})\left(\left|(\nabla - i\mathbf{A}_0)\,\psi\right|^2 + \frac{1}{2\varepsilon^2}\left(1 - |\psi|^2\right)^2\right)d\mathbf{x} \qquad (1.10)$$

where Ω denotes the platform of the film, $a(\mathbf{x})$ measures the relative thickness of the film and $\mathbf{A}_0$ is a prescribed vector potential due to the normal (to film) component of the applied field. The role of the Ginzburg-Landau parameter κ is assumed by the parameter $\varepsilon(\simeq 1/\kappa)$.

It has been proved that for fixed ε, the minimizers of the above problem, along with the prescribed vector potential $\mathbf{A}_0$ provide the leading order approximation to the solution of the three dimensional problem [CDG]. The creation and interaction of vortices based on the above model is connected with the prescribed magnetic potential. The number of vortices cannot be prescribed a priori, independently of $\mathbf{A}_0$. In order to simplify the analysis further, a simpler problem, in which the number of vortices is prescribed and the magnetic potential is ignored, can be studied. By rescaling the spatial variables, one may consider the minimization of the following functional:

$$\mathcal{F}_\epsilon^a(\psi) = \int_\Omega a(\mathbf{x})\left(|\nabla\psi|^2 + \frac{1}{2\varepsilon^2}\left(1 - |\psi|^2\right)^2\right)d\mathbf{x}. \qquad (1.11)$$

with the boundary condition:

$$\psi(\mathbf{x}) = g(\mathbf{x}) \quad \text{for} \quad \mathbf{x} \in \partial\Omega. \qquad (1.12)$$

Where g is smooth with $|g(\mathbf{x})| = 1$, $\mathbf{x} \in \partial\Omega$. We shall simply consider here the variational problem (1.11)–(1.12) with $a(\mathbf{x}) = 1$.

2. The Work of Bethuel-Brézis-Hélein and Others

Let Ω be a smooth, bounded, connected domain in $\mathbb{R}^2$. We consider the variational problem: minimize

$$E_\varepsilon(u) \equiv \int_\Omega \frac{1}{2}\left[|\nabla u|^2 + \frac{1}{2\varepsilon^2}\left(|u|^2 - 1\right)^2\right] dx, \qquad (2.1)$$

for $u \in H_g^1(\Omega, \mathbb{R}^2) = \{v \in H^1(\Omega, \mathbb{R}^2) : v|_{\partial\Omega} = g\}$. Consider the case that $\deg(g, \partial\Omega) \equiv d = 0$, see [BBH], and hence there is a smooth map $g^* : \Omega \to \mathbb{S}^1$ with $g^* = g$ on $\partial\Omega$. In particular,

$$\begin{aligned} \nu_\varepsilon &= \min\left\{E_\varepsilon(u) : u \in H_g^1(\Omega, \mathbb{R}^2)\right\} \\ &\leq C(\Omega, g) < \infty. \end{aligned}$$

We now give a different proof of the main result shown first in [BBH].

By a theorem of C. B. Morrey, there is a map $U_0 : \Omega \to \mathbb{S}^1$ which minimizes $\int_\Omega |\nabla U|^2 dx$ over the set $H_g^1(\Omega, \mathbb{S}^1) = \{U \in H^1(\Omega, \mathbb{S}^1) : U = g \text{ on } \partial\Omega\}$. Moreover, U_0 is smooth. When Ω is simply connected, a simple lifting argument shows that $U_0 = e^{i\phi_0}$. Here ϕ_0 is the harmonic extension of ϕ, $e^{i\phi} = g$ on $\partial\Omega$.

Here we want to show a uniform estimate for the minimizers U_ε of (2.1) for $0 < \varepsilon < 1$ under the hypothesis that $\deg(g, \partial\Omega) = 0$.

To do so , we note first that

$$E_\varepsilon(U_\varepsilon) \leq \int_\Omega |\nabla U_0|^2 dx, \qquad \text{for all } 0 < \varepsilon < 1. \qquad (2.2)$$

For any sequence $\varepsilon_i \to 0$, there is a subsequence of U_{ε_i} which converges weakly in H^1 and strongly in L^2 to some $U_* \in H^1(\Omega, \mathbb{S}^1)$. Moreover, $U_* = g$ on $\partial\Omega$, and $\int_\Omega |\nabla U_*|^2 dx \leq \int_\Omega |\nabla U_0|^2 dx$. By the minimizing property of U_0, we see U_* again is a minimizer of $\int_\Omega |\nabla U|^2 dx$ over $H_g^1(\Omega, \mathbb{S}^1)$. In particular U_* is smooth. Moreover, it follows from (2.2) that U_{ε_i} converges strongly to U_*. We therefore obtain the following

Lemma 1. *For any $\varepsilon_0 > 0$, there is an $r_0 > 0$ depending on $\partial\Omega$ and g such that if U_ε is a minimizer of (2.1) then*

$$\int_{\Omega \cap B(x, r_0)} \left[|\nabla U_\varepsilon|^2 + \frac{1}{2\varepsilon^2}\left(|U_\varepsilon|^2 - 1\right)^2\right] dx \leq \varepsilon_0 \qquad (2.3)$$

for all $x \in \bar\Omega$ provided $0 < \varepsilon \leq \varepsilon_*(r_0, \varepsilon_0)$.

Proof. Let F denote the set of energy minimizing maps over the set $H_g^1(\Omega, \mathbb{S}^1)$. Then it is easy to see that F is compact in $H^1(\Omega, \mathbb{S}^1)$. Moreover, by Morrey's theorem, one has for any $\varepsilon_0 > 0$, $U_* \in F$

$$\int_{\Omega \cap B(x, r)} |\nabla U_*|^2 dx \leq \varepsilon_0/2 \qquad (2.4)$$

for all $x \in \bar\Omega$ and $0 < r \leq r_0$ provided that r_0 is chosen to be suitably small.

Now we apply the convergence argument above to conclude that (2.3) is valid for all minimizers U_ε whenever $0 < \varepsilon \le \varepsilon_*$. Note that $(1/\varepsilon^2) \int_\Omega (|U_\varepsilon|^2 - 1)^2 \, dx \to 0$ as $\varepsilon \to 0^+$.

Theorem 1. *Let U_ε be a minimizer of (2.1) over the set $H_g^1(\Omega, \mathbb{C})$ with $\deg(g, \partial\Omega) = 0$. Then*

$$\sup_\Omega \left[|\nabla U_\varepsilon|^2 + \frac{1}{2\varepsilon^2} \left(|U_\varepsilon|^2 - 1 \right)^2 \right] \le C(g, \Omega) \tag{2.5}$$

for all $0 < \varepsilon < 1$.

Proof. It is obvious, by the maximum-principle, that $|U_\varepsilon| \le 1$ on Ω. Thus $|U_\varepsilon|(1 - |U_\varepsilon|^2) \, 1/\varepsilon^2 \le 1/\varepsilon^2 \le 1/\varepsilon_*^2$ whenever $\varepsilon \ge \varepsilon_*$. It follows that (2.5) is true whenever $\varepsilon \ge \varepsilon_*$.

For $0 < \varepsilon < \varepsilon_*$, we need the following lemmas.

Lemma 2. *Let $e_\varepsilon(u) = \frac{1}{2} \left[|\nabla u|^2 + \frac{1}{2\varepsilon^2} \left(|u|^2 - 1 \right)^2 \right]$, then*

$$\Delta \, e_\varepsilon(U_\varepsilon) \ge -3 \left[e_\varepsilon^2(U_\varepsilon) + e_\varepsilon(U_\varepsilon) \right] \ .$$

Proof. A direct computation yields the conclusion.

Lemma 3. *There is a positive number $\eta_0 \in (0, 1)$ such that, if $\int_{B_{r_0}(0)} e_\varepsilon(U_\varepsilon) \, dx \le \eta_0$, then*

$$\sup_{B_{r_0/2}(0)} e_\varepsilon(U_\varepsilon) \le C_1 \fint_{B_{r_0}(0)} e_\varepsilon(U_\varepsilon) \, dx \le \frac{C_1}{r_0^2} \, \eta_0 \ .$$

Proof. Look at the function $v(x) = (r_0 - |x|)^2 \, e_\varepsilon(U_\varepsilon)$, and set $K^2 = v(x_0) = \max_{B_{r_0}(0)} v(x), \sigma_0 = r_0 - |x_0|$.

We claim: if η_0 is small enough, then $K \le 1$. For otherwise, we may consider $V_\varepsilon(y) = U_\varepsilon \left(x_0 + \frac{\sigma_0}{K} y \right)$ defined on $B_K(0)$. It is easy to see that $e(V_\varepsilon) = \frac{\sigma_0^2}{K^2} e(U_\varepsilon)$, $e_\varepsilon(V_\varepsilon) \ge \frac{\sigma_0^2}{K^2} e_\varepsilon(U_\varepsilon)$, $e(V_\varepsilon) \le 4$ on $B_{K/2}(0)$ and $e_\varepsilon(V_\varepsilon)(0) = 1$.

Since $\Delta e_\varepsilon(V_\varepsilon) + 3 \left[e_\varepsilon^2(V_\varepsilon) + e_\varepsilon(V_\varepsilon) \right] \ge 0$, $\sup_{B_{1/4}} e_\varepsilon(V_\varepsilon) \le C_1 \fint_{B_{1/2}} e_\varepsilon(V_\varepsilon) \, dy = C_1 \int_{B_{\sigma_0/2K}(x_0)} e_\varepsilon(U_\varepsilon) \, dx \le C_1 \eta_0 < 1$, by choosing η_0 suitably small. Thus we obtain a contradiction. Hence $K \le 1$, in particular $e_\varepsilon(U_\varepsilon) \le r_0^{-2}$ on $B_{r_0/2}(0)$. The conclusion of Lemma 3 follows.

In the proof above, we have implicitly assumed that $B_{r_0}(0) \subset \Omega$. If $B_{r_0}(0) \cap \partial\Omega \ne \phi$, we refer to [CL] for the details.

Finally, we consider $\Psi_\varepsilon = \frac{1}{\varepsilon^2}(1 - |U_\varepsilon|^2) \ge 0$,

$$\begin{cases} 2\varepsilon^2 \Delta \Psi_\varepsilon - \Psi_\varepsilon \ge -4|\nabla U_\varepsilon|^2 \ge -C_1, & \text{for all} \quad 0 < \varepsilon \le \varepsilon_*. \\ \Psi_\varepsilon = 0 \quad \text{on} \quad \partial\Omega \ . \end{cases} \tag{2.6}$$

Let $x_0 \in \Omega$ such that $\Psi_\varepsilon(x_0) = \max_{x \in \bar\Omega} \Psi_\varepsilon(x) > 0$. Then $\Delta\Psi_\varepsilon(x_0) \le 0$, and thus $\Psi_\varepsilon(x_0) \le C_1$. The latter implies that

$$\|\Delta U_\varepsilon\|_{L^\infty(\Omega)} \le C_1 \quad \text{and} \quad U_\varepsilon \to U_0 \quad \text{in} \quad C^{1,\alpha}(\bar\Omega), \tag{2.7}$$

for $\alpha \in (0,1)$ follows.

Next we shall describe the main result in [BBH]. Here $\deg(g,\Omega) = d > 0$.

Theorem [BBH]. *Let $\varepsilon_n \downarrow 0$, and $\{U_{\varepsilon_n}\}$ be a sequence of minimizers of (2.1). Then, by taking subsequences if necessary, one has the following:*

(i) $U_{\varepsilon_n}(x) \longrightarrow U_*(x) = \prod_{j=1}^d \frac{x - a_j}{|x - a_j|} e^{i h_a(x)}, \quad in \quad C^{1,\alpha}(\bar\Omega\backslash\{a_1, \ldots, a_d\}),$
$\Delta h_a = 0 \quad in \ \Omega, \ U_* = g \ on \ \partial\Omega;$

(ii) *There is an $\varepsilon_0 > 0$ such that, for $0 < \varepsilon < \varepsilon_0$, U_ε has exactly d distinct zero's $a_1^\varepsilon, \ldots, a_d^\varepsilon$, and each zero is of degree 1. Moreover, the d-tuple point $a = (a_1, \ldots, a_d)$ is a global minimum of the renormalized energy $W(g, \Omega, b)$, $b \in \Omega^d$, (see next lecture for details concerning the renormalized energy);*

(iii) $\dfrac{(|U_{\varepsilon_n}|^2 - 1)^2}{\varepsilon_n^2} \longrightarrow 2\pi \sum_{j=1}^d \delta_{a_j}, \qquad \dfrac{|\nabla U_{\varepsilon_n}|^2}{\log \frac{1}{\varepsilon_n}} \longrightarrow 2\pi \sum_{j=1}^d \delta_{a_j},$
in the sense of distributions;

(iv) $E_\varepsilon(U_\varepsilon) = \pi d \log \dfrac{1}{\varepsilon} + \gamma d + \min_{b \in \Omega^d} W(g, \Omega, b) + o(1).$

Remarks (a) The above statements were shown in [BBH] under the additional assumption that Ω is star-shaped. The key conclusion following from this assumption is that the quantity

$$\int_\Omega \frac{1}{\varepsilon^2}\left(|U_\varepsilon|^2 - 1\right)^2 dx$$

is bounded by $C(g, \Omega)$. Using the approach in [St], one can drop this additional assumption. Indeed, the estimate

$$\int_\Omega \frac{1}{\varepsilon^2}\left(|U_\varepsilon|^2 - 1\right)^2 dx \le C(g, \Omega) \tag{2.8}$$

also follows from [St]. Later in [FP] an elegant approach showed also this estimate without using the star-shaped property of Ω.

(b) The proof of (iii) follows easily from (i) and the result of [BCP]. In fact, one can combine the proof of [BCP] and the classical result of E. Heinz to conclude that

$$Jac(U_{\varepsilon_n}) \longrightarrow \pi \sum_{j=1}^d \delta_{a_j}.$$

(c) Some more refined estimates due to P. Mironescu et al. were obtained recently. In particular, they showed

$$E_\varepsilon(U_\varepsilon) = \pi\, d \log \frac{1}{\varepsilon} + \gamma d + \min_{b \in \Omega^d} W(g,\, \Omega,\, b) \tag{2.9}$$

$$+ O(\varepsilon^\beta), \quad \text{for some} \quad \beta > 0\,.$$

Here γ is a universal positive constant.

(d) If Ω is star-shaped, it is also shown in [BBH] that results similar to the above are valid for arbitrary solutions, which are not necessarily minimizing. Without the additional assumption that Ω is star-shaped the conclusions fail in general. Under the assumption that the energy is bounded above by $\pi d \log \frac{1}{\varepsilon} + C$, for some C (independent of ε), we have more precise conclusions, cf. [L3]. We also constructed solutions which have both positive and negative degree zeros in [L3].

(e) Similar results for the full Ginzburg-Landau energy functionals (with magnetic field) have been obtained in [BR]. See also [DL] under a more physical boundary condition, and with an applied magnetic field.

(f) Let $a_\varepsilon \in \Omega$ be one of the zero's of U_ε. By Theorem [BBH], we may assume, for a sequence $\varepsilon_n \downarrow 0$, $a_{\varepsilon_n} \to a_0$, $U_{\varepsilon_n} \to U_*$ in $C^{1,\alpha}(B_{R_0}(a_0) \setminus \{a_0\})$. Moreover, $U_* = \frac{x - a_0}{|x - a_0|} e^{ih}$ in $B_{R_0}(a_0)$, for a smooth harmonic function h defined on $B_{R_0}(a_0)$.

By [BMR], one may assume $V_n(x) = U_{\varepsilon_n}(a_{\varepsilon_n} + \varepsilon_n x) \to F(x)$ in $C^k(B_R(0))$, $\forall R > 0$, $\forall k \geq 0$, where F satisfies

$$-\Delta F = \left(1 - |F|^2\right) F \quad \text{on} \quad \mathbb{R}^2\,, \quad \text{with} \quad F(0) = 0 \tag{2.10}$$

and
$$\int_{\mathbb{R}^2} \left(1 - |F|^2\right)^2 dx = 2\pi\,.$$

Moreover, F is locally energy minimizing in $\mathbb{R}^2$.

Entire solutions of (2.10) have the property (cf. [BMR]) $\int_{\mathbb{R}^2} \left(1 - |F|^2\right)^2 dx = 2\pi d^2$, for $d = 0, 1, 2, \ldots,$. Moreover, when $d < \infty$, $F\left(r\, e^{i\theta}\right) \simeq e^{i\, d(\theta + \theta_0)}$, as $r \to \infty$, for some $\theta_0 \in \mathbb{R}$. (cf. [S]).

It is still an open problem to classify all entire solutions of (2.10) with $\int_{\mathbb{R}^2} (1 - \|F\|^2)^2 dx < \infty$, even for the case that F is locally energy minimizing. In the latter case, one can show (cf. [S]) that $F\left(r e^{i\theta}\right) \simeq e^{i\,(\theta + \theta_0)}$, as $r \to \infty$. It was shown later by Shafrir, and also independently in personal communications of the author and H. Brézis that

$$\lim_{n \to \infty} \left\| U_{\varepsilon_n}(x) - F\left(\frac{x - a_{\varepsilon_n}}{\varepsilon_n}\right) e^{i\,(h(x) - h(a_0))} \right\|_{L^\infty\left(B_{R_0}(a_0)\right)} = 0\,. \tag{2.11}$$

It is an open problem, however, if (2.11) is true when the L^∞ norm is replaced by H^1-norm.

(g) One of the most challenging problem remaining to understand is the uniqueness of energy minimizers. Let us look at a simple situation. One considers

$$\begin{cases} -\Delta U_\varepsilon & = & \left(1 - |U_\varepsilon|^2\right) U_\varepsilon & \text{in} \quad B_1 \ , \\ U_\varepsilon(x) & \equiv & x \ , & x \in \partial B_1 \ . \end{cases} \tag{2.12}$$

There is a solution of the form $f_\varepsilon(r)\, e^{i\,\theta}$ such that $f_\varepsilon(r) \to 1$, as $\varepsilon \to 0^+$, for all $r > 0$. It is not known if $U_\varepsilon^0 \equiv f_\varepsilon(r)\, e^{i\,\theta}$ is the only solution of (2.12).

Recently, Lieb and Loss [LiL] and Mironescu [M] showed independently that U_ε^0 is a stable (and strictly stable) solution, and hence it is locally energy-minimizing. T. C. Lin [Lin] showed further that the linearized equation at U_ε^0 has exactly two small positive eigenvalues corresponding to the translation invariance of (2.10) and the other eigenvalues are strictly positive (independent of ε). There is also an interesting result shown in [CK] that the entire solutions F of (2.10) are of the form $F = f_\varepsilon(r)\, e^{i\,\theta}$ (up to translation and rotation) whenever $\int_{\mathbb{R}^2}(1 - |F|^2)^2\, dx < \infty$ and that either

$$\text{div } F = 0 \quad \text{or} \quad \text{curl } F = 0 \text{ in } \mathbb{R}^2 \ .$$

There is good reason to believe the solution of (2.12) is unique (for small ε). First of all, the Renormalized energy (defined in the next lecture) has a unique critical point, the origin, which is also the global minimum. Thus, by [BBH], any minimizer U_ε of (2.1) tends to $\frac{x}{|x|}$ as $\varepsilon \to 0^+$. Here we want to show the latter fact is true even for solutions of (2.12).

To see this, we note first, by Pohozaev's identity, that

$$\int_{\partial B_1} \left(\frac{\partial U_\varepsilon}{\partial \nu}\right)^2 + \int_{B_1} \frac{1}{\varepsilon^2}\left(1 - |U_\varepsilon|^2\right)^2 = 2\pi \ .$$

Hence, by the general convergence result in [BBH, Chapter X], one has $\frac{\left(|U_\varepsilon|^2 - 1\right)^2}{\varepsilon^2} \to 2\pi\,\delta_a$, for some $a \in B_1$. Moreover, a is a critical point of the renormalized energy $W(g, B, b)$, $b \in B_1$, where $g(x) = \frac{x}{|x|}$. A simple computation shows that $W(g, B, b) = -\log(1 - |b|^2)$, and thus $a = \underline{0}$. It, therefore, follows that $U_\varepsilon(x) \to \frac{x}{|x|}$ as $\varepsilon \to 0^+$.

Proof of Theorem [BBH] (Sketch of ideas).
We should sketch the ideas involved in the proof of the first statement (i). The statements in (iii) are easy consequences of the convergence result (i). The proof of the fact that the d-tuple point $a = (a_1, \ldots, a_d)$ is a global minimum of $W(g, \Omega, b)$, $b \in \Omega^d$ follows from the following:

1. A direct consequence of a comparison argument shows that

$$\min\left\{E_\varepsilon(U) : u \in H_g^1\left(\Omega, \mathbb{R}^2\right)\right\} = E_\varepsilon(U_\varepsilon)$$

$$\leq \pi d \log \frac{1}{\varepsilon} + d\gamma + \min_{b \in \Omega^d} W(g, \Omega, b) + o(1) \ .$$

Here $o(1) \to 0$ as $\varepsilon \to 0^+$;

2. The convergence result in part (i) and another comparison argument yield that

$$E_\varepsilon(U_\varepsilon) \geq \pi d \log \frac{1}{\varepsilon} + d\gamma + W(g, \Omega, a_\varepsilon) + o(1) \,,$$

where $a_\varepsilon = \left(a_\varepsilon^1, \ldots, a_\varepsilon^d\right)$ are d-tuple points consisting of zero's of U_ε. (cf. Lecture 2).

Thus the key step in proving Theorem [BBH] is to prove part (i). Now we sketch its proof. We shall simply assume Ω is star-shaped (this assumption is unnecessary, by [St],[FP]).

Step I. First, $E_\varepsilon(U_\varepsilon) \leq \pi d \log \frac{1}{\varepsilon} + C(g, \Omega)$ by a direct comparison. Next, the Maximum principle implies that $|U_\varepsilon|(x) \leq 1$, $x \in \Omega$. Then by a simple scaling (by looking at the equation satisfied by $U_\varepsilon(\varepsilon\, x)$) one deduces $|\nabla U_\varepsilon(x)| \leq \frac{C}{\varepsilon}$, $x \in \Omega$.

Also a Pohozaev-type identity yields:

$$\int_{\partial\Omega} \left|\frac{\partial U_\varepsilon}{\partial \nu}\right|^2 + \int_\Omega \frac{1}{\varepsilon^2} \left(|U_\varepsilon|^2 - 1\right)^2 \, dx \leq C(g, \Omega) \,.$$

Step II. One covers the set $S_\varepsilon = \{x \in \Omega : |U_\varepsilon(x)| \leq \frac{1}{2}\}$ by balls of radius ε. Thus, there are balls, $B_\varepsilon(x_j)$, $j = 1, \ldots, N_\varepsilon$, such that $S_\varepsilon \subseteq \bigcup_{j=1}^{N_\varepsilon} B_\varepsilon(x_j)$ and that $B_{\varepsilon/5}(x_j), j = 1, \ldots, N_\varepsilon$ are pairwise disjoint.

Since $|U_\varepsilon(x_j)| \leq \frac{1}{2}$, $|\nabla U_\varepsilon| \leq \frac{C}{\varepsilon}$, one sees that $\int_{B_{\varepsilon/5}(x_j)} \frac{1}{\varepsilon^2} \left(|U_\varepsilon|^2 - 1\right)^2 \, dx \geq C_0 > 0$, for some positive C_0, for each $j = 1, \ldots, N_\varepsilon$. This latter fact, and estimates in Step I imply that $N_\varepsilon \leq N(g, \Omega)$.

Step III. Let $U : A_{r,R} = \{x \in \mathbb{R}^2 : r \leq |x| \leq R\} \to \mathbb{R}^2$, $r \geq \varepsilon$, satisfy $|U| \geq \frac{1}{2}$ and $\int_{A_{r,R}} \frac{1}{\varepsilon^2} \left(|U|^2 - 1\right)^2 \, dx \leq K$. Then $\frac{1}{2} \int_{A_{r,R}} |\nabla U|^2 \, dx \geq \pi d^2 \log \frac{R}{r} - C(K, d)$, where $d = \deg\left(\widehat{U}, \partial B_\rho\right)$, $r \leq \rho \leq R$, $\widehat{U}(x) = \frac{U(x)}{|U|(x)}$.

Proof. We write $U = f\, e^{i\,(d\theta + \Psi(\rho, \theta))}$, then $|\nabla U|^2 \geq f^2 \left|\frac{d}{\rho} + \frac{\Psi_\theta}{\rho}\right|^2 \geq \frac{d^2}{\rho^2}\, f^2 + \frac{1}{4} \frac{|\Psi_\theta|^2}{\rho^2} + \frac{2f^2}{\rho^2}\, d \cdot \Psi_\theta$.

Thus

$$\frac{1}{2} \int_{A_{r,R}} |\nabla U|^2 \, dx \geq \frac{d^2}{2} \int_{A_{r,R}} \frac{dx}{\rho^2} + \frac{d^2}{2} \int_{A_{r,R}} \frac{f^2 - 1}{\rho^2} \, dx$$
$$+ 2d \int_{A_{r,R}} \frac{(f^2 - 1)}{\rho^2}\, \Psi_\theta \, dx + \frac{1}{4} \int_{A_{r,R}} \frac{|\Psi_\theta|^2}{\rho^2} \, dx$$

$$\geq d^2 \pi \log \frac{R}{r} - \frac{d^2}{2} \left[\int_{A_{r,R}} \left(f^2 - 1 \right)^2 dx \right]^{1/2} \left(\int_r^R \frac{2\pi}{\rho^3} d\rho \right)^{1/2}$$

$$- \frac{1}{4} \int \frac{\Psi_\theta^2}{\rho^2} dx - 4d^2 \int \frac{(f^2 - 1)^2}{\rho^2} dx + \frac{1}{4} \int \frac{\Psi_\theta^2}{\rho^2} dx$$

$$\geq d^2 \pi \log \frac{R}{r} - \frac{d^2}{2} \varepsilon K^{1/2} \left(4\pi \varepsilon^{-2} \right)^{1/2} - C(d, K)$$

$$\geq d^2 \pi \log \frac{R}{r} - C(K, d) .$$

Step IV. By a simple grouping and induction argument, one can use the results in Step I, Step II and Step III to conclude that, as $\varepsilon \to 0$, the sets $\{x_j\}_{j=1}^N$ converge (in Hausdorff distance) to $\{a_j\}_{j=1}^d$, for some d distinct points $a_1, \ldots, a_d$ in Ω. Moreover, $\deg \left(\widehat{U}_\varepsilon, \partial B_\rho(a_j) \right) = 1$, for each j, and $\rho > 0$ (suitably small) whenever ε is sufficiently small. Also $\displaystyle\int_{\Omega \setminus \bigcup_{j=1}^d B_\rho(a_j)} |\nabla U_\varepsilon|^2 dx \leq C(\rho)$. The final estimate will imply the conclusion of part (i) in Theorem [BBH]. We refer to [St] and [BBH] for details. $\qquad\square$

By employing the same arguments as above, one can show the following somewhat more general statement:

Lemma 4. *Let U be a minimizer of the functional (2.1) in a Lipschitz domain Ω with $U = g$ on $\partial\Omega$. Suppose that*

$$\int_{\partial\Omega} \left[\left| \frac{\partial g}{\partial \nu} \right|^2 + \frac{1}{2\varepsilon^2} \left(|g|^2 - 1 \right)^2 \right] \leq K ,$$

for a constant K. Then, for all sufficiently small $\varepsilon > 0$ (depending on K and Ω) we have

$$E_\varepsilon(U) \leq C(K, \Omega)$$

whenever $\deg(g, \partial\Omega) = 0$, and

$$E_\varepsilon(U) \geq \pi |d| \log \frac{1}{\varepsilon} - C(K, \Omega) .$$

if $\deg(g, \partial\Omega) = d \neq 0$.

Lemma 5. *With the same hypothesis as in Lemma 4, suppose $\deg(g, \Omega) = 0$. Then $|U(x)| \geq \frac{3}{4}$ in Ω whenever $0 < \varepsilon \leq \varepsilon(K, \Omega)$, for some positive $\varepsilon(K, \Omega)$. In general, let $v \in H^1(\Omega, \mathbb{R}^2)$ with $v = g$ on $\partial\Omega$ and $|\nabla v| \leq \frac{C}{\varepsilon}$. Then $E_\varepsilon(v) \geq \min \{E_\varepsilon(u) : u|_{\partial\Omega} = g\} + c_0(K, \Omega)$ for some positive $c_0(K, \Omega)$ whenever $|v(\underline{0})| \leq \frac{1}{2}$ for a point $\underline{0} \in \Omega$.*

The proof of Lemma 5 follows from Theorem 1 and its proof.

3. Some Generalizations

To end this lecture, we would like to present two general results proved in [L]. Both of these results will play an important role in the next two lectures.

For this purpose, we introduce the following:

Definition 1. *Let $\Omega \subset \mathbb{R}^2$ be a smooth, bounded domain, and let $g : \partial\Omega \to \mathbb{S}^1$ be a smooth map of degree $d > 0$. We say a map $U : \Omega \to \mathbb{R}^2$ belongs to the class $S(c_0, K, \varepsilon, g, \Omega)$ if $U \in H_g^1(\Omega, \mathbb{R}^2)$ and*

(i) $E_\varepsilon(U) \leq \pi d \log \dfrac{1}{\varepsilon} + K$,

(ii) *for $x_0 \in \Omega$ with $|U(x_0)| \leq \frac{1}{2}$, then $|U(x)| \leq \frac{3}{4}$ whenever $x \in \Omega$, $|x - x_0| \leq c_0\,\varepsilon$.*

We have the following structure theorem concerning $U \in S(c_0, K, \varepsilon, g, \Omega)$. The proof given in [L] is somewhat more complicated, but it is useful for other purposes. Here we shall give a simplified proof of the following:

Structure Theorem. *There are two positive numbers ε_0, α_0 depending only on c_0, K, g and Ω such that, for any $0 < \varepsilon < \varepsilon_0$, $u \in S(c_0, K, \varepsilon, g, \Omega)$, there are N_ε disjoint balls B_j of radius ε^{α_j}, $j = 1, \ldots, N_\varepsilon$ with the following properties:*

(i) $\alpha_0 \leq \alpha_j \leq 1$, *for $j = 1, \ldots, N_\varepsilon$, and $N_\varepsilon \leq N(c_0, K, g, \Omega)$.*

(ii) *The set $\left\{ x \in \Omega : |u(x)| \leq \frac{1}{2} \right\}$ is contained in $\Omega \cap \left(\bigcup\limits_{j=1}^{N_\varepsilon} B_j \right)$.*

(iii) *The estimates $\varepsilon^{\alpha_j} \int_{\partial(B_j \cap \Omega)} e_\varepsilon(u) \leq C(c_0, \alpha_0, K, g, \Omega)$, $j = 1, \ldots, N_\varepsilon$, are valid. In particular, the degrees $d_j = \deg\left(\frac{u}{|u|}, \partial(B_j \cap \Omega) \right)$ are well-defined. Here $e_\varepsilon(u) = |\nabla u|^2 + \frac{1}{2\varepsilon^2}\left(|u|^2 - 1 \right)^2$.*

(iv) *There are exactly d balls, say $B_1, \ldots, B_d$ such that the corresponding degrees d_j, $j = 1, \ldots, d$, are not zero. Moreover, each d_j equals 1, for $j = 1, \ldots, d$. Suppose $x_1, \ldots, x_d$ be centers of balls $B_1, \ldots, B_d$, then*

$$\min\{ \operatorname{dist}(x_j, \partial\Omega), |x_i - x_j|, i \neq j, i, j = 1, \ldots, d \} \geq \delta_* = \delta_*(c_0, K, g, \Omega) > 0 \ .$$

(v) *If $B_j \cap \Omega$ is scaled by a factor of size $\simeq \varepsilon^{-\alpha_j}$, the resulting domain has diameter one and is uniformly Lipschitz (independent of j and ε).*

Proof. Let $u \in S(c_0, K, \varepsilon, g, \Omega)$, then

$$E_\varepsilon(u) = \frac{1}{2} \int_\Omega \left[|\nabla u|^2 + \frac{1}{2\varepsilon^2}\left(|u|^2 - 1 \right)^2 \right] dx \leq \pi d \log \frac{1}{\varepsilon} + K \ .$$

Let $\underline{u}$ be a minimizer of the energy functional

$$\underline{E}_\varepsilon(v) \equiv \frac{1}{2} \int_\Omega \left[|\nabla v|^2 + \frac{1}{4\varepsilon^2}\left(|v|^2 - 1 \right)^2 \right] dx \ , \quad v \in H_g^1(\Omega, \mathbb{R}^2) \ .$$

By Lemma 4,

$$\pi d \log \frac{1}{\varepsilon} - C(\Omega, g) \leq \underline{E}_\varepsilon(u) \leq \underline{E}_\varepsilon(u).$$

Thus

$$E_\varepsilon(u) - \underline{E}_\varepsilon(u) = \frac{1}{8\varepsilon^2} \int_\Omega \left(1 - |u|^2\right)^2 \, dx \leq K + C(\Omega, g).$$

Using the last fact, and noting that if $x_0 \in \{x \in \Omega : |u(x)| \leq \frac{1}{2}\}$, then

$$\int_{B_{\varepsilon/2}(x_0)} \frac{(|u|^2 - 1)^2}{\varepsilon^2} \, dx \geq C_1 > 0,$$

as $u \in S(c_0, K, \varepsilon, g, \Omega)$, we see, from the proof of Theorem [BBH] that the set $\{x \in \Omega : |u(x)| \leq \frac{1}{2}\}$ can be covered by $B_\varepsilon(x_j)$, $j = 1, \ldots, M_\varepsilon$, $M_\varepsilon \leq N(c_0, K, g, \Omega)$.

Now we want to find (for any $\varepsilon > 0$ suitably small) at most N_ε balls $\overline{B}_j$, of radius $\varepsilon^{\overline{\alpha}_j}$, covering the same set and satisfying the additional conditions:

(I) $\overline{\alpha}_j \in \left[\alpha_0, \dfrac{1}{4}\right]$, for $j = 1, \ldots, N_\varepsilon \leq N(c_0, K, g, \Omega)$.

Here α_0 is a positive constant which may depend on N_ε.

(II) The set $\{x \in \Omega : |u(x)| \leq \frac{1}{2}\}$ is contained in $\Omega \cap \bigcup_{j=1}^{N_\varepsilon} \overline{B}_j$; and the balls $\varepsilon^{-\overline{\alpha}_j/3} \, \overline{B}_j$ (with same center as $\overline{B}_j$ and radius $\varepsilon^{-\overline{\alpha}_j/3}$) are pairwise disjoint, for $j = 1, 2, \ldots, N_\varepsilon$.

To prove these two properties (I) and (II), we need the following:

Covering Lemma. *Let $B_1, B_2, \ldots, B_N$ be N balls in R^2, each with radius no larger than ε^α, for some $\alpha \in (0, 1/4)$. Then there are a positive number α_0 (depending only on α and N), and balls $\overline{B}_j$ of radius $\varepsilon^{\overline{\alpha}_j}$, for $j = 1, \ldots, N_\varepsilon \leq N$, such that (I) and (II) above are valid provided that ε is sufficiently small.*

Proof. Let $A = \bigcup_{j=1}^{N} B_j$. We are going to prove the covering Lemma by induction on the number of connected components of A. If A is connected, then we simply take $\overline{\alpha}_1 = \frac{\alpha}{3}$, and a ball $\overline{B}_1$ of radius $\varepsilon^{\overline{\alpha}_1} \geq 2N\varepsilon^\alpha$ (this inequality will be valid whenever ε is suitably small) such that $A \subset \overline{B}_1$. The conclusion of the covering Lemma follows automatically.

Suppose that the conclusions of the covering lemma are true whenever the number of the connected components of A is $1 \leq k \leq N - 1$. Moreover, these $\overline{\alpha}_j$'s satisfy $\overline{\alpha}_j \geq \frac{\alpha}{3^k}$, for each j.

We want to show the covering lemma is true when the number of the connected components of A is $k + 1 \leq N$, and that each $\bar{\alpha}_j$ in the lemma can be chosen to be not less than $\frac{\alpha}{3^{k+1}}$ whenever ε is small enough.

For this purpose, we let $A_1, \ldots, A_{k+1}$ be connected components of A. Without loss of the generality, we may assume that the diameter of A is larger than $3(k + 1)\varepsilon^{\frac{2\alpha}{3^{k+1}}}$. For, otherwise, we may simply choose a ball B of radius $\leq \varepsilon^{\frac{2\alpha}{3^{k+1}}}$ that covers A entirely (whenever ε is small enough), and then the conclusion of the covering lemma is obvious.

Now we let $x', x'' \in A$ be such that $|x' - x''| = $ diameter of $A \geq 3(k + 1)\varepsilon^{\frac{2\alpha}{3^{k+1}}}$. We may find a $\rho_0 \in \left(0, 3(k+1)\varepsilon^{\frac{2\alpha}{3^{k+1}}}\right)$ such that the boundary $\partial B_r(x')$ of the ball $B_r(x')$ will not intersect any of A_j's, for $j = 1, \ldots, k + 1$, and for any $r \in \left[\rho_0 - \varepsilon^{\frac{2\alpha}{3^{k+1}}}, \rho_0 + \varepsilon^{\frac{2\alpha}{3^{k+1}}}\right]$. Then it is obvious that $A \cap B_{\rho_0}(x') = A'$, and $A'' = A \sim A'$, each contains some of $A_1, \ldots A_{k+1}$. We may apply the induction step to both A' and A'' to conclude that $A = A' \cup A''$ can be covered by at most N balls $\overline{B}_j$ of radius $\varepsilon^{\bar{\alpha}_j}$, $\bar{\alpha}_j \geq \frac{\alpha}{3^k}$. Now since dist $(A', \partial B_{\rho_0}(x')) \geq \varepsilon^{\frac{2\alpha}{3^{k+1}}}$, and since dist $(A'', \partial B_{\rho_0}(x')) \geq \varepsilon^{\frac{2\alpha}{3^{k+1}}}$, the conclusions of the covering lemma follows. This completes the induction argument. □

Now we can apply Fubini's Theorem again, to find balls B_j (with the same center as $\overline{B}_j$) of radius ε^{α_j}, $\alpha_j \in [\bar{\alpha}_j/3, , \bar{\alpha}_j]$ such that (i), (ii) and (iii) of the Structure Theorem are valid. Part (iv) and (v) follow from the same proof as in Step IV of Theorem [BBH]. □

We shall call $x_1, \ldots, x_d$ in the statement (iv) of the Structure Theorem the essential zero's of u. It is then clear that x_j's are well-defined up to errors which are not larger than $2\varepsilon^{\alpha_0}$. The latter statement means: if y_j, $j = 1, \ldots, d$, are another possible choice of essential zero's of u as those x_j's in the statement (iv) above, then

$$|x_j - y_j| \leq 2\varepsilon^{\alpha_0}, \quad \text{for} \quad j = 1, \ldots, d .$$

Compactness Theorem. *Let $U_\varepsilon \in S(c_0, K, \varepsilon, g, \Omega)$. Then, for any $\varepsilon_n \downarrow 0$, there is a subsequence of $\{U_{\varepsilon_n}\}$ that converges to a map of the form $\prod_{j=1}^{d} \frac{x-b_j}{|x-b_j|} e^{i h(x)}$ weakly in $H^1_{\text{loc}}\left(\bar{\Omega} \setminus \{b_1, \ldots, b_d\}\right)$. Moreover, $\|h\|_{H^1_{\text{loc}}} \leq C(c_0, K, g, \Omega)$ and $W(g, \Omega, b) \leq C(c_0, K, g, \Omega)$. The last statement implies, in particular, that*

$$\min\left\{ |b_i - b_j|, \quad \text{dist } (b_i, \partial\Omega), \ i \neq j, \ i, j = 1, \ldots, d \right\} \geq \delta_*(c_0, K, g, \Omega) .$$

Proof. For $U_\varepsilon \in S(c_0, K, \varepsilon, g, \Omega)$, we let B_j, $j = 1, \ldots, N_\varepsilon$ be balls as in the Structure Theorem. We replace U_ε on each B_j by $\tilde{U}_\varepsilon$ which minimizes $\int_{B_j} e_\varepsilon(v) \, dx$ with $v = U_\varepsilon$ on ∂B_j, for $j = d + 1, \ldots, N_\varepsilon$. Thus, in particular, $|\tilde{U}_\varepsilon| \geq \frac{3}{4}$ on B_j, for $j = d + 1, \ldots, N_\varepsilon$ (see Lemma 5). We will denote the resulting map defined on Ω by $\tilde{U}_\varepsilon$.

Let $\delta \in (\varepsilon^{\alpha_0}, \delta_*)$, where α_0, δ_* are given in the Structure Theorem, and suppose that the set $\bigcup_{j=1}^{d} \partial B_\delta(x_j)$ does not intersect the set $\bigcup_{j=1+d}^{N_\varepsilon} B_j$. Note that the latter assumption holds outside of a set of $\delta \in (\varepsilon^{\alpha_0}, \delta_*)$ whose measure is smaller than $N_\varepsilon \, \varepsilon^{\alpha_0}$.

For such a δ, we have

$$
\begin{aligned}
\pi \, d \log \frac{\delta}{\varepsilon} - C(\lambda, K) &\le \sum_{j=1}^{d} \int_{B_\delta(x_j)} e_\varepsilon(\tilde{U}_\varepsilon) \, dx \\
&\le \int_{A_\varepsilon} e_\varepsilon(\tilde{U}_\varepsilon) \, dx + C, \qquad\qquad (*) \\
\text{where} \qquad A_\varepsilon &= \textstyle\bigcup_{j=1}^{d} B_\delta(x_j) \big\backslash \bigcup_{j \ge d+1} B_j .
\end{aligned}
$$

Here the first inequality is true since (by lemma 4)

$$
\int_{B_j} e_\varepsilon(\tilde{U}_\varepsilon) \, dx = \int_{B_j} e_\varepsilon(U_\varepsilon) \, dx \ge \pi \, d \log \frac{1}{\varepsilon^{\alpha_j}} - C
$$

and

$$
\begin{aligned}
\int_{B_\delta(x_j)\backslash B_j} e_\varepsilon(\tilde{U}_\varepsilon) \, dx &\ge \frac{1}{2} \int_{B_\delta(x_j)\backslash B_j} |\nabla \tilde{U}_\varepsilon|^2 \, dx \\
&\ge \pi \, d \log \frac{\delta}{\varepsilon^{\alpha_j}} - C, \quad \text{for } j = 1, 2, \dots, d,
\end{aligned}
$$

see Step III in the proof of Theorem [BBH]. In the second inequality of $(*)$ we have used Lemma 5 for $\tilde{U}_\varepsilon$ on each B_j, $d + 1 \le j \le N_\varepsilon$.

Since $u_\varepsilon \in S(c_0, K, \varepsilon, g, \Omega)$, we deduce from $(*)$ that

$$
\int_{\Omega_\delta u \left(\bigcup_{j=d+1}^{N_\varepsilon} B_j \right)} e_\varepsilon(u_\varepsilon) \, dx \le C + \pi \, d \log \frac{1}{\delta}
$$

where $\Omega_\delta = \Omega \setminus \cup_{j=1}^{d} B_\delta(x_j)$.

Now, for a sequence of $\varepsilon_n \downarrow 0$, we may assume, without loss of generality, that $x_j \to b_j$ as $\varepsilon_n \downarrow 0$. Note x_j may also depend on ε. We may also assume, by taking a subsequence if necessary, that $u_{\varepsilon_n}(x) \to u^*(x)$ weakly in $H^1_{\text{loc}} (\overline{\Omega} \setminus \{b_1, \dots, b_d\})$ and strongly in $L^2(\Omega)$. The conclusion that

$$
u^*(x) = \prod_{j=1}^{d} \frac{x - b_j}{|x - b_j|} \, e^{i \, h(x)}
$$

with $h(x) \in H^1(\Omega)$ follows. $\square$

Lecture 2. The Role of the Renormalized Energy

1. Renormalized Energy

We start with the definition of the renormalized energy for the functionals (2.1) of Lecture 1. Let Ω be a bounded smooth domain in $\mathbb{R}^2$, and let $g : \partial\Omega \to \mathbb{S}^1$ be a smooth map of degree $d > 0$. To any d distinct points $a_1, \ldots, a_d$ in Ω, we can associate a canonical harmonic map $u_a : \overline{\Omega}\setminus\{a_1, \ldots, a_d\} \to \mathbb{S}^1$, which is smooth on $\overline{\Omega}\setminus\{a_1, \ldots, a_d\}$ and such that $u_a = g$ on $\partial\Omega$, the degrees of $u_a : \partial B_\rho(a_j) \to \mathbb{S}^1$ are all equal to 1, for $j = 1, \ldots, d$, and for all small $\rho > 0$.

One can write

$$u_a(x) = \prod_{j=1}^{d} \frac{x - a_j}{|x - a_j|} e^{ih_a}(x), \tag{1.1}$$

for some harmonic function h_a defined on Ω. The value of h_a on $\partial\Omega$ is uniquely determined (mod $\cdot 2\pi$) by the requirement $u_a = g$ on $\partial\Omega$.

A simple computation shows that

$$\frac{1}{2} \int_{\Omega\setminus \bigcup_{j=1}^{d} B_\rho(a_j)} |\nabla u_a|^2(x)dx = \pi d \log \frac{1}{\rho} + W(a_1, \ldots, a_d) \tag{1.2}$$

$$+ O(\rho), \quad \text{and} \quad \rho \to 0^+.$$

The function $W(a)$, $a \in \Omega^d$, which depends on g and Ω, is called the renormalized energy, see [BBH, Chapter I].

It is easy to check that $W(a)$ possesses the following property:

$$W(a) = +\infty \tag{1.3}$$

if either one of $a_i \in \partial\Omega$, for some $i \in \{1, \ldots, d\}$ or $a_i = a_j$, for some $i \neq j$. Otherwise $W(a)$ is locally analytic.

A rather important connection between the above renormalized energy and the Ginzburg-Landau energy functionals

$$E_\varepsilon(u) = \frac{1}{2} \int_\Omega \left[|\nabla u|^2 + \frac{1}{2\varepsilon^2} \left(|u|^2 - 1 \right)^2 \right] dx, \quad 0 < \varepsilon < 1, \tag{1.4}$$

defined on $H_g^1(\Omega) = \{u \in H^1(\Omega, \mathbb{R}^2) : u|_{\partial\Omega} = g\}$, is described in the previous lecture.

Here we are interested in the following question (cf. [BBH], open problem #6]):

Q. Let $a = (a_1, \ldots, a_d)$ be a nondegenerate critical point of the renormalized energy $W(\cdot)$. Is there a sequence of critical points u_{ε_n} of E_{ε_n}, $n = 1, 2, \ldots$, such that $\varepsilon_n \to 0$, and that $u_{\varepsilon_n}(x) \to u_a(x)$ in $C^{1,\alpha}(\overline{\Omega}\setminus\{a_1, \ldots, a_d\})$ as $n \to \infty$?

In [L], the author obtained a positive answer to the above question provided that a is a nondegenerate local minimum of $W(\cdot)$ on Ω^d. In fact, the case a is a degenerate local minimum of $W(\cdot)$ was also treated in [L]. The main result of

[L] was later improved by del Pino and Felmer, [DF]. Here we shall give another simple proof, based on some arguments in [L], of their improved result which can be stated as follows.

Theorem A. *Let K be a open subset with compact closure in $\Omega^d \backslash \{b \in \Omega^d: b_i = b_j,$ for some $i \neq j\}$. Let $a \in K$ be such that $\min_{\partial K} W > W(a) = \min_K W$. Then there is an $\varepsilon_0 > 0$ depending on g, Ω and K such that there is a family of local minimizers, $u_\varepsilon, 0 < \varepsilon \leq \varepsilon_0$, of E_ε with the following property: for any sequence $\varepsilon_n \to 0$, one may find a subsequence of $\{u_{\varepsilon_n}\}$ converging to u_b, $u_b(x) = \prod_{j=1}^{d} \frac{x-b_j}{|x-b_j|} e^{ih_b(x)}$, the canonical harmonic map associated with the point $b = (b_1, \ldots, b_d)$, for some point $b \in K_a$. Here $K_a = \{b \in K, W(b) = W(a)\}$.*

The main goal of this lecture is to establish various minimax solutions of the Euler-Lagrange equations associated with energy functionals E_ε, for small positive ε's. As a particular consequence we shall prove the following.

Theorem B. *Let $a = (a_1, \ldots, a_d)$ be a nondegenerate critical point of the renormalized energy $W(\cdot)$. Then there is an $\varepsilon_0 > 0$ depending only on g, Ω (and maybe also the point a) such that one may construct a family of critical points, $u_\varepsilon, 0 < \varepsilon \leq \varepsilon_0$, of E_ε with the property that, $u_\varepsilon \to u_a$, in $C_{\text{loc}}^{1,\alpha}(\overline{\Omega} \backslash \{a_1, \ldots, a_d\})$, as $\varepsilon \to 0^+$. Here u_a is given in (1.1).*

Theorem B, therefore, gives an affirmative answer to the question posed by Bethuel-Brézis-Hélein. Recently, we learned from the announcement [AB] that Almeida and Bethuel had used topological arguments to construct some non-minimal solutions of the Ginzburg-Landau equations.

Both the proof of existence of mountain-pass type critical points in [L] and that of existence of nonminimal solutions in [AB] used global information on the level sets of the renormalized energy $W(\cdot)$. The key point in establishing of Theorem B is to localize the nontrivial topological information of level sets of $W(\cdot)$ near a critical point in Ω^d, and to transfer it to the functionals E_ε on $H_g^1(\Omega)$. It still remains as an interesting open problem to calculate the Morse indices of these saddle points.

There are two important implications from the results stated above. The first one is that it reduces a variational problem in infinite dimensions, for which the numerical computations are rather unstable for ε small, to a finite dimensional problem. The second is that it may shed some light into the prediction of Abrikosov on the lattice structures of vortices in type II superconductors. Of course, in the latter situation, one also has to look at the applied field. See [BR] and [DL] for some related discussions. Here we simply describe a result in [DL] concerning the renormalized energy when there is an applied field.

With proper scaling, we focus on the following form of the G-L functional

$$\mathcal{G}(\psi, \mathbf{A}) = \int_\Omega \left(\frac{1}{4\epsilon^2}(1 - |\psi|^2)^2 + \frac{1}{2}|(\nabla - i\mathbf{A})\psi|^2 + \frac{1}{2}|\text{curl}\,\mathbf{A} - \mathbf{H}|^2 \right) dx$$

In this nondimensionalization, one may view ϵ as proportional to $\frac{1}{\kappa}$ and $\mathbf{H}$ as proportional to κ times the (nondimensionalized) applied field. For $n = 2$, let $\Omega \in R^2$ be a bounded Lipschitz domain, $\mathbf{H} = H_0$ a constant field.

Following the discussion in [BBH, BR], we now formulate the renormalized energy: let

$$\psi = e^{i\phi_b(\mathbf{x})} = \prod_{j=1}^{d} \frac{\mathbf{x} - \mathbf{b}_j}{|\mathbf{x} - \mathbf{b}_j|} e^{ih}$$

for some points $\mathbf{b} = (\mathbf{b}_1, \mathbf{b}_2, ..., \mathbf{b}_d) \in \Omega^d$ and $\frac{\partial \phi_b}{\partial \mathbf{n}} = 0$ on $\partial\Omega$. Let

$$B_\rho = \cup_{j=1}^{d} B_\rho(\mathbf{b}_j) \ .$$

Choose the gauge div $\mathbf{A} = 0$ in Ω and $\mathbf{A} \cdot \mathbf{n} = 0$ on $\partial\Omega$. We may define ζ, such that

$$\mathbf{A} = \nabla^\perp \zeta \text{ in } \Omega \ ,$$
$$\zeta = 0 \text{ on } \partial\Omega \ .$$

Now, consider

$$\begin{aligned}
\mathcal{G}_\rho &= \int_{\Omega \backslash B_\rho} \left(\frac{1}{2} |\nabla \phi_b - \nabla^\perp \zeta|^2 + |\Delta\zeta - H_0|^2 \right) d\Omega \\
&= \int_{\Omega \backslash B_\rho} \left(\frac{1}{2} |\nabla \phi_b|^2 + \frac{1}{2} |\nabla\zeta|^2 - \nabla \phi_b \cdot \nabla^\perp \zeta + \frac{1}{2} |\Delta\zeta - H_0|^2 \right) d\Omega \\
&= d\pi \log \frac{1}{\rho} + W_\Omega(\mathbf{b}, H_0) + O(\rho)
\end{aligned}$$

Note

$$\begin{aligned}
-\int_{\Omega \backslash B_\rho} \nabla \phi_b \cdot \nabla^\perp \zeta d\Omega &= \int_{\Omega \backslash B_\rho} \text{div} \, (\zeta \cdot \nabla^\perp \phi_b) \, d\Omega \\
&= \sum_{j=1}^{d} 2\pi\zeta(\mathbf{b}_j) + O(\rho)
\end{aligned}$$

So,

$$\begin{aligned}
W_\Omega(\mathbf{b}, H_0) &= \int_{\Omega \backslash B_\rho} \frac{1}{2} |\nabla \phi_b|^2 \, d\Omega - d\pi \log \frac{1}{\rho} + \sum_{j=1}^{d} 2\pi\zeta(\mathbf{b}_j) \\
&\quad + \int_{\Omega \backslash B_\rho} \frac{1}{2} \left(|\nabla\zeta|^2 + |\Delta\zeta - H_0|^2 \right) d\Omega + O(\rho) \\
&= \int_{\Omega \backslash B_\rho} \frac{1}{2} |\nabla \phi_b|^2 \, d\Omega - d\pi \log \frac{1}{\rho} + 2\pi \sum_{j=1}^{d} \zeta(\mathbf{b}_j) \\
&\quad + \int_{\Omega \backslash B_\rho} \frac{1}{2} \left(|\nabla\zeta|^2 + |\Delta\zeta|^2 \right) d\Omega + \frac{1}{2} H_0^2 |\Omega| - \int_{\Omega \backslash B_\rho} H_0 \Delta\zeta \, d\Omega + O(\rho)
\end{aligned}$$

Minimizing the term involving ϕ_b we see that ϕ_b is a multi-valued harmonic function on $\Omega \setminus \{\mathbf{b}_1, \mathbf{b}_2, \ldots, \mathbf{b}_d\}$,

$$\int_{\Omega \setminus B_\rho} \frac{1}{2} |\nabla \phi_b|^2 \, d\Omega - d\pi \log \frac{1}{\rho} = g_\Omega(\mathbf{b}) + O(\rho)$$

Minimizing the terms involving ζ, we can choose ζ to satisfy

$$-\Delta^2 \zeta + \Delta \zeta = 2\pi \sum_{j=1}^{d} \delta_{\mathbf{b}_j} \quad \text{in } \Omega$$
$$\zeta = 0 \quad \text{on } \partial\Omega$$
$$\Delta\zeta = H_0 \quad \text{on } \partial\Omega$$

So,

$$2\pi \sum_{j=1}^{d} \zeta(\mathbf{b}_j) = -\int_\Omega \left(|\nabla\zeta|^2 + |\Delta\zeta|^2 \right) d\Omega + H_0 \int_{\partial\Omega} \frac{\partial\zeta}{\partial n} \, d\Gamma \ .$$

Since H_0 is a constant, we get

$$\int_{\Omega \setminus B_\rho} H_0 \, \Delta\zeta \, d\Omega = H_0 \int_{\partial\Omega} \frac{\partial\zeta}{\partial n} \, d\Gamma + H_0 \, O(\rho)$$

Thus,

$$W_\Omega(\mathbf{b}, H_0) = \frac{1}{2} H_0^2 |\Omega| - \frac{1}{2} \int_\Omega \left(|\nabla\zeta|^2 + |\Delta\zeta|^2 \right) d\Omega + H_0 O(\rho) + O(\rho) + g_\Omega(\mathbf{b})$$

Now, let us define $\zeta = \zeta_b + \zeta_{H_0}$ where

$$-\Delta^2 \zeta_b + \Delta \zeta_b = 2\pi \sum_{j=1}^{d} \delta_{\mathbf{b}_j} \quad \text{in } \Omega$$
$$\zeta = 0 \quad \text{on } \partial\Omega$$
$$\Delta\zeta = 0 \quad \text{on } \partial\Omega$$

and $\zeta_{H_0} = H_0\zeta_1$ with

$$-\Delta^2 \zeta_1 + \Delta \zeta_1 = 0 \quad \text{in } \Omega$$
$$\zeta_1 = 0 \quad \text{on } \partial\Omega$$
$$\Delta\zeta_1 = 1 \quad \text{on } \partial\Omega$$

Then

$$\int_\Omega |\nabla\zeta|^2 \, d\Omega = H_0^2 \int_\Omega |\nabla\zeta_1|^2 \, d\Omega + \int_\Omega |\nabla\zeta_b|^2 \, d\Omega - 2H_0 \int_\Omega \Delta\zeta_1 \, \zeta_b \, d\Omega$$
$$\int_\Omega |\Delta\zeta|^2 \, d\Omega = H_0^2 \int_\Omega |\Delta\zeta_1|^2 \, d\Omega + \int_\Omega |\Delta\zeta_b|^2 \, d\Omega + 2H_0 \int_\Omega \Delta\zeta_1 \, \Delta\zeta_b \, d\Omega$$

and

$$\int_\Omega \Delta\zeta_1 \left(\Delta\zeta_b - \zeta_b\right) d\Omega \;=\; \int_\Omega \zeta_1 \Delta\left(\Delta\zeta_b - \zeta_b\right) d\Omega$$

$$= \; -2\pi \sum_{j=1}^{d} \zeta_1(\mathbf{b}_j)$$

So,

$$\int_\Omega \left(|\nabla\zeta|^2 + |\Delta\zeta|^2\right) d\Omega \;=\; H_0^2 \int_\Omega \left(|\nabla\zeta_1|^2 + |\Delta\zeta_1|^2\right) d\Omega$$

$$+ \int_\Omega \left(|\nabla\zeta_b|^2 + |\Delta\zeta_b|^2\right) d\Omega - 2\pi H_0 \sum_{j=1}^{d} \zeta_1(\mathbf{b}_j)$$

Therefore,

$$W_\Omega(\mathbf{b}, H_0) = \frac{1}{2}\, H_0^2\, C(\Omega) + \pi\, H_0 \sum_{j=1}^{d} \zeta_1(\mathbf{b}_j) + \tilde{g}_\Omega(\mathbf{b}) + O(\rho) \qquad (1.5)$$

where $C(\Omega)$ is a constant and $\tilde{g}_\Omega(\mathbf{b})$ has the property

$$\tilde{g}_\Omega(\mathbf{b}) = \begin{cases} +\infty & \mathbf{b}_i = \mathbf{b}_j \text{ for some } i \neq j\,, \\ -\infty & \mathbf{b} \in \partial\Omega^d \end{cases}$$

and otherwise, it is a smooth function in Ω^d.

It is then easy to show that $W_\Omega(\mathbf{b}, H_0)$ has a local minimum inside Ω^d whenever $H_0 \geq H_0(d, \Omega)$. Moreover, one can establish a result similar to Theorem A.

2. A Technical Result

We construct a natural embedding of $\Omega^d \backslash \{b \in \Omega^d\colon\; b_i = b_j, \text{ for some } i \neq j\}$. Let $b_1, b_2, \ldots, b_d$ be d distinct points in Ω. Then we can choose a small ρ (ρ can be chosen to be uniform for any compact subset of $A \equiv \Omega^d \backslash \{b \in \Omega\colon b_i = b_j \text{ for some } i \neq j\}$), so that there is a map $w_{\varepsilon,b}$ satisfying

$$E_\varepsilon(w_{\varepsilon,b}) = \pi d \log \frac{1}{\varepsilon} + \gamma d + W(b) + 0(\rho) + o(1).$$

Here $o(1)$ is a quantity which goes to zero as $\varepsilon \to 0$ (uniformly on any compact subset of A), $0(\rho) \to 0$ as $\rho \to 0$ (uniformly on any compact subset of A), and γ is a universal constant. The map $w_{\varepsilon,b}$ can be chosen so that $w_{\varepsilon,b}(x) = u_b(x)$ for $|x - b_j| \geq \rho$, $u_0(x - b_j)$ for $|x - b_j| \leq \frac{\rho}{2}$, and that on $B_\rho(b_j) \backslash B_{\rho/2}(b_j)$, $w_{\varepsilon,b}$ is the canonical harmonic map with admissible boundary conditions, for $j = 1, \ldots, d,$. (u_0 is a minimizer of $\int_{B_{\rho/2}(0)} e_\varepsilon(u)$ with the boundary condition $\frac{x}{|x|}$ on $\partial B_{\rho/2}(0)$). It is easy to see, for a given compact subset K of A, one may find such a $\rho > 0$, such that for any $b \in K$, one may construct $w_{\varepsilon,b}$ as above. Then $b \in K \to w_{\varepsilon,b}$ gives a continuous embedding of K into $H_g^1(\Omega)$.

Next, we want to construct a projection from the mapping class $S(\varepsilon, c_0, K, g, \Omega)$ to its essential zeros which lie in the set $A \subset \Omega^d$.

Lemma 1. *Let D be a compact topological submanifold with boundary in $\mathbb{R}^{2d}$, and let $h \colon D \to H_g^1(\Omega)$ be a continuous map such that $h(D) \subset S(\varepsilon, c_0, K, g, \Omega)$. Then there is a constant $\lambda = \lambda(c_0, K, g, \Omega)$, and a continuous map $\Pi \colon h(D) \to W_\lambda = \{b \in \Omega^d \colon W(b) \leq \lambda\}$ whenever $0 < \varepsilon \leq \varepsilon_0$. Moreover, for any $p \in D$, $\Pi(h(p))$ is at most $4\varepsilon_0^{\alpha_0}$ distant from $b(p)$ where $b(p) = (b_1(p), \dots, b_d(p))$, and $b_j(p)$'s are essential zero's of $h(p) \in S(\varepsilon, c_0, K, g, \Omega)$.*

Proof. For a point $p \in D$, we let $b_1(p), \dots, b_d(p)$ be a choice of essential zero's of the map $h(p) \in S(\varepsilon, c_0, K, g, \Omega)$. As $0 < \varepsilon \leq \varepsilon_0$, the points $b_1(p), \dots, b_d(p)$ are defined up to errors of, at most, $2\varepsilon^{\alpha_0}$ (cf. Structure Theorem). Since $h \colon D \to H_g^1(\Omega)$ is continuous, one may find an open ball V_p around p, such that, for any $q \in V_p$, $b_1(p), \dots, b_d(p)$, can be also viewed as essential zero's of the map $h(q)$. Therefore, we have an open cover $\{V_p, p \in D\}$ of D. Since D is compact, we may find a finite subcover $\{V_{p_j}, j = 1, \dots, m\}$. Moreover, by the Besicovitch covering lemma, we may also assume that $\sum_{j=1}^{m} \chi_{V_{p_j}}(p) \leq c(d), \forall p \in D$.

Let $\{\phi_j\}$ be a partition of unity subordinate to the cover $\{V_{p_j}\}$. That is each ϕ_j is smooth and nonnegative, with support $\phi_j \subset V_{p_j h}$, and $\sum_{j=1}^{m} \phi_j = 1$. We let

$$\Pi(h(p)) = \sum_{j=1}^{m} B_j \phi_j(p), \qquad B_j = (b_1(p_j), \dots, b_d(p_j)).$$

Then Π is obviously continuous on $h(D)$. Moreover, since each $p \in D$, there are at most $c(d)$ number of V_{p_j}'s such that $p \in V_{p_j}$. By our construction, if $p \in V_{p_j} \cap V_{p_k}$, then $|B_k - B_j| \leq 4\varepsilon_0^{\alpha_0}$. The conclusion of the Lemma 7 follows from the latter estimate and the Compactness Theorem. $\qquad\square$

3. Proof of Theorem A

First let us consider the case that a is a nondegenerate minimum of $W(\cdot)$. We may assume the set K in the Theorem A is given by $\prod_{j=1}^{d} B_{R_0}(a_j) \subseteq A \equiv \Omega^d \backslash \{b \in \Omega^d \colon b_i = b_j, \text{ for some } i \neq j\}$, and so that $K_a = \{a\}$. The reason we can do so is that K contains a set of form $\prod_{j=1}^{d} B_{R_0}(a_j)$, for some R_0, and the conclusion of the Theorem A will not be affected.

We introduce

$$V = \{u \in H^1_g(\Omega): |u| \geq \frac{3}{4} \text{ on } \Omega \setminus \bigcup_{j=1}^{d} B_{R_0}(a_j), \text{ and}$$

$$\text{degree}\left(\frac{u}{|u|}, \partial B_{R_0}(a_j)\right) = 1, \text{ for } j = 1, 2, \ldots, d\}$$

and

$$V_a = \{u \in V: u \in C(\overline{\Omega}) \text{ and } u \text{ has exactly one}$$

$$\text{zero in each } B_{R_0}(a_j), \text{ for } j = 1, 2, \ldots, d\}.$$

It is clear that V is weakly closed in $H^1_g(\Omega)$. One can also show that V is connected, though we will not need this. Therefore the $\inf_{u \in V} E_\varepsilon(u)$ is achieved.

We *claim:*

$\inf_{u \in V_a} E_\varepsilon(u)$ is also achieved and that is equal to $\inf_{u \in V} E_\varepsilon(u)$.

To show the claim, we let $u_n \in V_a$ be a minimizing sequence. Let $x^n_1, \ldots, x^n_d$ be zero's of u_n. Since $E_\varepsilon(u_n) \leq \pi d \log \frac{1}{\varepsilon} + W(a) + \gamma d + 0_\varepsilon(1)$ is obviously true by the explicitly construction in Section 1, we may find balls $B_{\varepsilon^{\alpha_j}}(x^n_j)$, for $j = 1, 2, \ldots, d$ (and for a given n), such that $\varepsilon^{\alpha_j} \int_{\partial B_{\varepsilon^{\alpha_j}}(x^n_j)} e_\varepsilon(u_n) \leq C(\alpha_0, a, g, \Omega)$ (*cf. Structure Theorem*). Moreover, as $u_n \in V_a$, $\det\left(\frac{u_n}{|u_n|}, \partial B_{\varepsilon^{\alpha_j}}(x^n_j)\right) = 1$. Replace u_n by $\tilde{u}_n$ on the set $\Omega \setminus \bigcup_{j=1}^{d} B_{\varepsilon^{\alpha_j}}(x^n_j)$ so that $u_n = \tilde{u}_n$ on $\partial\left[\Omega \setminus \bigcup_{j=1}^{d} B_{\varepsilon^{\alpha_j}}(x^n_j)\right]$, and that $\tilde{u}_n$ minimizes $\int_{\Omega \setminus \bigcup_{j=1}^{d} B_{\varepsilon^{\alpha_j}}(x^n_j)} e_\varepsilon(u) dx$.

Without loss of generality, we may assume that $x^n_j \to \bar{a}_j$ as $n \to \infty$. Then by the proof of [L, Theorem A], one has that $\tilde{u}_n$ converges as $\varepsilon \to 0$, $n \to \infty$, to the map

$$u_{\bar{a}} = \prod_{j=1}^{d} \frac{x - \bar{a}_j}{|x - \bar{a}_j|} e^{ih_{\bar{a}}(x)} \text{ in } C^{1,\alpha}_{\text{loc}}(\overline{\Omega} \setminus \{\bar{a}_1, \ldots, \bar{a}_d\}).$$

Moreover, the proof there also showed that

$$E_\varepsilon(u_n) \geq \pi d \log \frac{1}{\varepsilon} + \gamma d + W(\bar{a}) + 0_\varepsilon(1).$$

Since $\bar{a} \in \prod_{j=1}^{d} \overline{B}_{R_0}(a_j)$, and a is a nondegenerate minimum of W in $\prod_{j=1}^{d} B_{R_0}(a_j)$, one sees that, for sufficiently small ε, the limit $\bar{a}_j$ of x^n_j as $n \to \infty$ (for the fixed ε) satisfies $|a_j - \bar{a}_j| < \frac{R_0}{4}$. Thus, for n large enough, one may assume $B_{\varepsilon^{\alpha_j}}(x^n_j) \subseteq B_{R_0/2}(a_j)$.

The new minimizing sequence $\hat{u}_n$ which is given by u_n on each $B_{\varepsilon^{\alpha_j}}(x_j^n)$ and $\tilde{u}_n$ on $\Omega \backslash \bigcup\limits_{j=1}^{d} B_{\varepsilon^{\alpha_j}}(x_j^n)$ is also in V_a. As $\hat{u}_n \to u_{\bar{a}}$, as $n \to \infty$ and $\varepsilon \to 0$, we may assume $\hat{u}_n|_{\partial B_{R_0}(a_j)}$ and $u_{\bar{a}}|_{\partial B_{R_0}(a_j)}$ are $C^{1,\alpha}$ close for small ε, and large n. Let u_n^* be the map which is equal to $\hat{u}_n$ on $\Omega \backslash \bigcup\limits_{j=1}^{d} B_{R_0}(a_j)$ and which minimizes $\int_{B_{R_0}(a_j)} e_\varepsilon(u)$ on each $B_{R_0}(a_j)$ (with respect to its Dirichlet boundary condition). From the fact that $u_{\bar{a}}|_{\partial B_{R_0}(a_j)}$ is a diffeomorphism from $\partial B_{R_0}(a_j)$ to S^1 (for a suitable small R_0, and we may assume R_0 is so chosen at the beginning), u_n^* has a unique zero in $B_{R_0}(a_j)$ of degree 1 (cf. [BCP]).

Finally the minimizing sequence $\{u_n^*\}$, (for a sufficiently small ε, but fixed) of E_ε on V_a is also equicontinuous for all n. Moreover the limit, as $n \to \infty$, of u_n^* is obviously also in V_a (cf. [BCP]).

We have, therefore, proved that the $\inf\limits_{V_a} E_\varepsilon(u)$ is achieved. It is obvious that $\inf\limits_{V} E_\varepsilon(u) \leq \inf\limits_{V_a} E_\varepsilon(u)$. This completes the proof of the claim.

Next we let u_ε be a minimizer of E_ε in V. Let $x_j \in B_{R_0}(a_j)$ be such that $|u_\varepsilon(x_j)| \leq \frac{1}{2}$ in the Lebesgue sense, for $j = 1, \ldots, d$. As before, we may find balls $B_{\varepsilon^{\alpha_j}}$, $\alpha_0 \leq \alpha_j \leq 1$, such that

$$\varepsilon^{\alpha_j} \int_{\partial B_{\varepsilon^{\alpha_j}}(x_j)} e_\varepsilon(u_\varepsilon) \leq C(\alpha_0, a, g, \Omega).$$

We claim that $\deg\left(\frac{u_\varepsilon}{|u_\varepsilon|}, \partial B_{\varepsilon^{\alpha_j}}(x_j)\right) = 1$, for all $j = 1, 2, \ldots, d$, whenever ε is small enough.

As in the proof of [[L], Theorem A], it suffices to show

$$\text{degree}\left(\frac{u_\varepsilon}{|u_\varepsilon|}, \partial B_{\varepsilon^{\alpha_j}}(x_j)\right) \neq 0.$$

Indeed, when ε is small enough, one has $|u_\varepsilon| > \frac{4}{5}$ on each $\partial B_{\varepsilon^{\alpha_j}}(x_j)$, $j = 1, \ldots, d$. Assume, for some j, $\deg\left(\frac{u_\varepsilon}{|u_\varepsilon|}, \partial B_{\varepsilon^{\alpha_j}}(x_j)\right) = 0$. Then we may replace u_ε inside the ball $B_{\varepsilon^{\alpha_j}}(x_j)$ by an energy minimizer $\hat{u}_\varepsilon$ with the same Dirichlet boundary condition. Such a minimizer $\hat{u}_\varepsilon$ satisfies $|\hat{u}_\varepsilon| \geq \frac{4}{5}$ inside $B_{\varepsilon^{\alpha_j}}(x_j)$. Thus if we let u_ε^* be the map coinciding with u_ε on $\Omega \backslash B_{\varepsilon^{\alpha_j}}(x_j)$ and equal to $\hat{u}_\varepsilon$ on $B_{\varepsilon^{\alpha_j}}(x_j)$, then $u_\varepsilon^* \in V$. Moreover $E_\varepsilon(u_\varepsilon^*) < E_\varepsilon(u_\varepsilon)$ as $|u_\varepsilon(x_j)| \leq \frac{1}{2}$ in the Lebesgue sense. This contradicts to the minimality of u_ε.

Therefore we may assume each $\deg\left(\frac{u_\varepsilon}{|u_\varepsilon|}, \partial B_{\varepsilon^{\alpha_j}}(x_j)\right) = 1$. Now we follow the first part of the argument to show as $\varepsilon \to 0$, $x_j \to a_j$, and $u_\varepsilon \to u_a$ in $C_{loc}^{1,\alpha}(\overline{\Omega} \backslash \{a_1, \ldots, a_d\})$. in $C_{loc}^{1,\alpha}(\overline{\Omega} \backslash \{a_1, \ldots, a_d\})$. Then $u_\varepsilon \in V_a$ follows from [BCP] again. It is easy to see from the above argument that if u_ε is a minimizer in either V_a or V of the energy E_ε, then $|u_\varepsilon| > \frac{3}{4}$ in $\Omega \backslash \bigcup\limits_{j=1}^{d} B_{R_0}(a_j)$. Therefore u_ε is a local minimizer of E_ε in $H_g^1(\Omega)$. This completes the proof of Theorem A in the case that a is a nondegenerate minimum of W.

In the general case, we may define function spaces

$$V(K) = \{u \in H_g^1(\Omega): |u| \geq \frac{3}{4} \text{ on } \Omega \setminus \bigcup_{j=1}^d B_{R_0}(b_j),$$

$$\deg\left(\frac{u}{|u|}, \partial B_{R_0}(b_j)\right) = 1, \text{ for } j = 1, \ldots, d,$$

$$\text{for some point } b = (b_1, \ldots, b_d) \in \overline{K}\}.$$

Here $R_0 > 0$ has to be chosen suitably small (which may depend on K).

Again one can show $V(K)$ is weakly closed and $\inf_{V(K)} E_\varepsilon(u) = E_\varepsilon(u_\varepsilon)$, for some $u_\varepsilon \in V(K)$. One can also define

$$V_c(K) = \{u \in V(K): u \text{ is continuous on } \overline{\Omega} \text{ and}$$

$$\text{there is a point } b \in K \text{ such that } u$$

$$\text{vanishes only at } b_1, b_2, \ldots, b_d\}.$$

One can use the exactly same proof as above to show that

$$\inf_{V_c(K)} E_\varepsilon(u) = \inf_{V(K)} E_\varepsilon(u).$$

Moreover, if u_ε is a minimizer of E_ε in either $V_c(K)$ or $V(K)$, then u_ε vanishes exactly at points $(a_1^\varepsilon, \ldots, a_d^\varepsilon) \to K_a$ as $\varepsilon \to 0^+$. Also $|u_\varepsilon| > \frac{4}{5}$ on $\Omega \setminus \bigcup_{j=1}^d B_{R_0/2}(a_j^\varepsilon)$, and thus u_ε's are local minimizers of E_ε over $H_g^1(\Omega)$.

This completes the proof of Theorem A.

4. Proof of Theorem B

We consider a nondegenerate critical point $a = (a_1, \ldots, a_d)$ of the renormalized energy $W(\cdot)$ defined on Ω^d. That is, the function W satisfies the following at the point a:

$$\nabla W(a) = 0 \text{ and } \nabla^2 W(a) \text{ has no zero eigenvalues.} \tag{4.1}$$

Thus we may find a $2d$ dimensional ball $B_{\delta_0}(a)$ centered at a and of radius δ_0, and a local orthonormal coordinate system $(x_1, \ldots, x_{2d})$ on $B_{\delta_0}(a)$ such that, in this coordinate system, a becomes the origin $(x_1, \ldots, x_{2d}) = (0, \ldots, 0)$, and that

$$W(x) = -\sum_{i=1}^k \lambda_i x_i^2 + \sum_{i=k+1}^{2d} \lambda_i x_i^2 + W(a) + 0(|x|^2), \tag{4.2}$$

for some positive numbers $\lambda_1, \ldots, \lambda_{2d}$, and some k, $0 \leq k \leq 2d$.

Note that the case $k = 0$ means a is a local nondegenerate minimum of $W(\cdot)$. For this case we have already established Theorem A in the previous section. Thus we may assume $k \geq 1$.

If $k = 2d$, then a is a local nondegenerate maximum of $W(\cdot)$. One may choose a suitable $\delta_0 > 0$ so that

$$-\delta_1 = \max_{b \in \partial B_{\delta_0}(a)} (W(b) - W(a)), \qquad \delta_1 > 0.$$

Moreover, let $D = B_{\delta_0}(a)$, and let $S = \{b \in \Omega^d: W(b) \geq W(a) - \delta_1/2\}$. Then ∂D and S are homotopically linked in Ω^d. That is, $\partial D \cap S = \phi$, and for any $h \in C(D, \Omega^d)$ with $h|_{\partial D} = id_{\partial D}$, $h(D) \cap S \neq \phi$.

In the process of proving Theorem B, we have to localize the linking property. In the case $k = 2d$, we see D and S are homotopically linked in any $B_{\delta_0+\sigma}(a) \subseteq \Omega^d$, for $\sigma > 0$. Moreover, one can easily verify that $\max W(h(D)) \geq W(a)$, for any $h \in C(D, B_{\delta_0+\sigma}(a))$ with $h|_{\partial D} = id_{\partial D}$.

Suppose now $1 \leq k \leq 2d - 1$. We set

$$D = \{x \in \mathbb{R}^{2d}: |x| < \delta_0, x_i = 0, \text{ for } i \geq k+1\},$$

and let $D^\perp = \{x \in \mathbb{R}^{2d}: |x| < \delta_0, x_i = 0, \text{ for } i \leq k\}$, where $(x_1, \ldots, x_{2d})$ is the coordinate system chosen so that (4.2) is true. Then

$$\delta_1 = \min\{\min_{x \in \partial D} (W(a) - W(x)), \min_{\partial D^\perp} (W(x) - W(a))\}$$

is positive whenever δ_0 is suitably small. Next we introduce a diffeomorphism h from the open ball $\{x \in \mathbb{R}^{2d}: |x| < \delta_0 - \frac{\sigma_0}{4}\}$ to $\mathbb{S}^{2d} \backslash \{N\}$. Here $\mathbb{S}^{2d}$ is the standard $2d$-dimensional sphere, and N is the north pole. We also assume h maps $\{x \in \mathbb{R}^{2d}: \delta_0 - \sigma_0/4 \leq |x| \leq \delta_0\}$ to N so that h becomes continuous from the closed ball $\{x \in \mathbb{R}^{2d}: |x| \leq \delta_0\}$ to $\mathbb{S}^{2d}$. We note that the image of $\overline{D}$ under h becomes a k-dimensional sphere in $\mathbb{S}^{2d}$.

We choose $\sigma_0 > 0$ so small that on the set $\{x \in \mathbb{R}^{2d}: \text{dist}(x, \partial D^\perp) \leq 2\sigma_0\}$ the function W is bounded from below by $W(a) - 2\delta_1/3$. We also note that under the map h, the image of the set $Y = \{x \in D^\perp: |x| = \delta_0 - \sigma_0\}$ is a $2d - k - 1$ dimensional sphere. Moreover, these two spheres $h(Y)$ and $h(\overline{D})$ are linked in $\mathbb{S}^{2d}$.

We shall now embed the k-dimensional ball $\overline{D}$, $k \geq 1$, into $S(\varepsilon, c_0, K, g, \Omega)$ as in Section 1. Thus, for each p, we have a map $w_p \in S(\varepsilon, c_0, K, g, \Omega)$ such that

$$E_\varepsilon(\omega_p) = \pi d \log \frac{1}{\varepsilon} + \gamma d + W(p) + o(1), \qquad (4.3)$$

and that

$$|\nabla w_p|(x) \leq \frac{c_1}{\varepsilon}, |\nabla w_p(x) - \nabla w_p(y)| \leq \frac{c_1|x - y|}{\varepsilon^2}, \text{ for } x, y \in \overline{\Omega}. \qquad (4.4)$$

where c_1 is a constant depending only on g, Ω and the point a, c_0 may be chosen so that it will depend only on c_1. The constant K in the definition of the class $S(\varepsilon, c_0, K, g, \Omega)$ can be chosen to be

$$K = \max_{b \in B_{\delta_0 + \sigma_0}(a)} W(b). \tag{4.5}$$

Moreover, the quantity $o(1)$ goes to zero uniformly on $\overline{D}$ as $\varepsilon \to 0^+$.

We remark that the embedding $I: p \in D \to w_p \in S(\varepsilon, c_0, K, g, \Omega) \subset H_g^1(\Omega)$ is continuous, and that w_p vanishes exactly at d points $p_1, p_2, \dots, p_d$ so that $p = (p_1, \dots, p_d)$. From now on we shall always assume ε to be so small that the quantity $o(1)$ in (4.3) is bounded by $\delta_1/4$ in absolute value.

Let $u_\varepsilon(t, p, \cdot)$ be the unique solution of the problem:

$$\frac{\partial u}{\partial t} = \Delta u + \frac{1}{\varepsilon^2}(1 - |u|^2)u \text{ in } \Omega \times R_+ \tag{4.6}$$

with initial and boundary conditions:

$$u_\varepsilon(0, p, x) = w_p(x), \qquad x \in \Omega, \tag{4.7}$$
$$u_\varepsilon(t, p, x) = g(x), \quad \text{for } x \in \partial\Omega, \quad t \geq 0. \tag{4.8}$$

It is then obvious that each $u_\varepsilon(t, p, \cdot) \in S(\varepsilon, c_0, K, g, \Omega)$. Moreover, the map $p \in \overline{D} \to u_\varepsilon(t, p, \cdot) \in S(\varepsilon, c_0, K, g, \Omega)$ is continuous (with respect to the H^1 norm metric on $H_g^1(\Omega)$) for each fixed $t \geq 0$. Indeed, the latter follows from the unique solvability of (4.6)–(4.8).

From the energy decreasing property of the flow (4.6) and our choice of the initial data w_p, we deduce, from (4.3), that

$$E_\varepsilon(u_\varepsilon(t, p, \cdot)) \leq \pi d \log \frac{1}{\varepsilon} + \gamma d + W(p) + 0(1) \tag{4.9}$$

$$\leq \pi d \log \frac{1}{\varepsilon} + \gamma d + W(a) + \delta_1/4.$$

Hence we may apply both structural and compactness theorem in Section 1 to all maps $u_\varepsilon(t, p, \cdot)$, for $t \geq 0$, $p \in \overline{D}$. In particular, it follows, when $\varepsilon \leq \varepsilon_0$, that the essential zeros of $u_\varepsilon(t, p, \cdot)$, say $b_1(t, p), \dots, b_d(t, p)$ are well-defined (up to possible errors that are not greater than $4\varepsilon^{\alpha_0}$). Moreover, the point $b(t, p) = (b_1(t, p), \dots, b_d(t, p))$ cannot lie in the set $\{x \in \mathbb{R}^{2d}: \text{dist}(x, \partial D^\perp) \leq 2\sigma_0\}$ for, otherwise, one deduce from the proof of [[L], Theorem A] that $E_\varepsilon(u_\varepsilon(t, p, \cdot) \geq \pi d \log \frac{1}{\varepsilon} + \gamma d + W(b(t, p)) + \frac{2\delta_1}{3} + 0(1) \geq \pi d \log \frac{1}{\varepsilon} + \gamma d + W(a) + \frac{\delta_1}{3}$. The latter contradicts to (4.9).

Next we observe, from [[L2], Theorem 4.5], that for any $t, t' \in R_+$ with $|t - t'| \leq 1$,

$$|b(t, p) - b(t', p)| \leq 8\varepsilon^{\alpha_0} + 0_\varepsilon(1). \tag{4.10}$$

Here $0_\varepsilon(1)$ is again a quantity which goes to zero (uniformly independent of t and t') as $\varepsilon \to 0$ where $b(t', p) = (b_1(t', p), \dots, b_d(t', p))$, and $b_j(t', p)$, for $j = 1, \dots, d$

are essential zeros of $u_\varepsilon(t', p, \cdot)$ (well-defined up to a possible error of ε^{α_0}). Therefore we may choose ε to be sufficiently small that the right-hand side is not larger than $\frac{\sigma_0}{8}$.

For $t_n = n$, $n = 1, 2, \ldots$, we may define continuous map Π_n: $\{u_\varepsilon(n, p, \cdot), p \in D\} \to W_K = \{b \in \Omega^d: W(b) \le K + \gamma d\}$ as in Lemma 1. Moreover $|\Pi_n(u_\varepsilon(n, p, \cdot) - \Pi_{n+1}(u_\varepsilon(n+1, p, \cdot))| \le \sigma_0/8$, and $\Pi_n(u_\varepsilon(n, p, \cdot))$ is at most $4\varepsilon^{\alpha_0}$ distance away from essential zeros of $u_\varepsilon(n, p, \cdot)$.

Now we define a continuous map, f: $\overline{D} \times \overline{R}_+ \to \Omega^d$ as follows:

$$\begin{aligned} f(p, 0) &= p, \qquad \forall \in \overline{D}, \\ f(p, n) &= \Pi_n(u_\varepsilon(n, p, \cdot)), \qquad \forall p \in \overline{D}, n = 1, 2, \ldots, \\ f(p, t) &= (t - k) f(p, k+1) + (k+1-t) f(p, k). \end{aligned}$$

Note that $f(p, t)$ is at most $\frac{\sigma_0}{4}$ distance away from essential zeros of $u_\varepsilon(p, t, \cdot)$. We also note that, by the energy decreasing property $f(p, t) \notin B_{\delta_0 - \sigma_0/4}(a)$ for all $t \ge 0$ and $p \in \partial D$. (Note $p \in \partial D$ achieves the minimum for $W(\cdot)$ on $\overline{B}_{\delta_0}(a)$).

We already noted at the beginning of this section that $h(Y)$ and $h(D)$ are linked in $\mathbb{S}^{2d}$. One may also extend h outside the ball $\{x \in \mathbb{R}^{2d}: |x| \le \delta_0\}$ by simply setting $h(x) = h\left(\frac{\delta_0 x}{|x|}\right)$, for $|x| > \delta_0$.

Now we consider $h(Y)$ and $h_t(D)$, $t \ge 0$, where $h_t = h \circ f(\cdot, t)$. It is clear $h_0 = h$ and h_t is a family of continuous maps from D to $\mathbb{S}^{2d}$ (continuous in t). Moreover $h_t(\partial D) = N$, for each $t \ge 0$. Finally, since $f(p, t) \notin Y$, we see that $h(Y) \cap h_t(D) = \phi$ for each $t \ge 0$. Therefore $h(Y)$ and $h_t(D)$ are linked in $\mathbb{S}^{2d}$ for each $t \ge 0$. From the latter fact, one concludes that for $n = 1, 2, \ldots$, there is a $p_n \in \overline{D}$ such that $f(p_n, n) \in D^\perp$. Then, it follows again from the proof of Theorem A in [L] that,

$$E_\varepsilon(u_\varepsilon(n, p_n, \cdot)) \ge \pi d \log \frac{1}{\varepsilon} + \gamma d + W(a) + 0_\varepsilon(1), \text{ for } n = 1, 2, \ldots. \qquad (4.11)$$

Since $E_\varepsilon(u_\varepsilon(n, p_n, \cdot)) \le E_\varepsilon(u_\varepsilon(o, p_n, \cdot))$, and by our choice of the initial data $w_p(\cdot)$, $p \in \overline{D}$, one obtains that $|p_n - a| \le \delta_\varepsilon$, for $n = 1, 2, \ldots$. Here δ_ε is a quantity which goes to zero when $\varepsilon \to 0^+$. Without loss of generality, we may assume $\lim_{n \to \infty} p_n = p$ exist. (By taking a subsequence if necessary; here ε is fixed.)

Then, on the one hand, one has

$$E_\varepsilon(u_\varepsilon(t, p, \cdot)) \le E_\varepsilon(u_\varepsilon(0, p, \cdot)) \le \pi d \log \frac{1}{\varepsilon} + \gamma d + W(a) + 0_\varepsilon(1), \qquad (4.12)$$

and, on the other hand, by the continuity of the map $p \to u_\varepsilon(t, p, \cdot)$, for each fixed p,

$$E_\varepsilon(u_\varepsilon(t, p, \cdot)) = \lim_{n \to \infty} E_\varepsilon(u_\varepsilon(t, p_n, \cdot)) \ge \overline{\lim_{n \to \infty}} E_\varepsilon(u_\varepsilon(n, p_n, \cdot)). \qquad (4.13)$$

Here we used the energy decreasing property of the flow (4.6)–(4.8) for each p_n.

From (4.12), (4.13) and (4.11), we have, for this p, that

$$\left| E_\varepsilon(u_\varepsilon(t,p,\cdot)) - \left(\pi d \log \frac{1}{\varepsilon} + W(a) + \gamma d \right) \right| \leq \eta_\varepsilon \to 0, \text{ as } \varepsilon \to 0^+, \text{ for any } t \geq 0.$$
(4.14)

Moreover, for this particular p, and for a fixed if small ε, one may take the time limit to obtain $u_\varepsilon(\infty, p, \cdot) = \lim_{t\to\infty} u_\varepsilon(t, p, \cdot)$ so that $E_\varepsilon(u_\varepsilon(\infty, p, \cdot)) = \pi d \log \frac{1}{\varepsilon} + W(a) + \gamma d + 0_\varepsilon(1)$. Finally, we can apply Corollary 5.5 of [L2] to conclude that $u_\varepsilon(x) = u_\varepsilon(\infty, p, x)$ has the property that

$$\lim_{\varepsilon \to 0^+} u_\varepsilon(x) = \prod_{j=1}^{d} \frac{x - a_j}{|x - a_j|} e^{ih_a(x)} \text{ in } C_{loc}^{1,\alpha}(\overline{\Omega}\backslash\{a_1, \ldots, a_d\}).$$

Here we have also used the property that a is the only critical point in $B_{\delta_0}(a)$ of $W(\cdot)$.

Remark 1. It is not hard to see that the above proof of Theorem B can also be applied to some cases where a may be a degenerate critical point of $W(\cdot)$. But, on the other hand, certain nontrivial topological information on the level sets of $W(\cdot)$ near the critical point a is necessary for the above arguments.

Remark 2. In the above proof, we have implicitly assumed that $1 \leq k \leq 2d - 1$. When $k = 2d$, the set Y will be empty. But when $k = 2d$, we follow the above proof to show that $h_t(D)$ is always the whole sphere S^{2d}. Thus, for $n = 1, 2, \ldots$, one may always find a $p_n \in D$ such that $f(n, p_n) = a$. From the latter fact, one immediately obtains the same conclusion that $|p_n - a| \leq \delta_\varepsilon$. Let $p_{n_i} \to p$, as $n_i \to \infty$ (here ε is fixed), then

$$\left| E_\varepsilon(u_\varepsilon(t,p,\cdot)) - \left(\pi d \log \frac{1}{\varepsilon} + W(a) + \gamma d \right) \right| \leq \delta_\varepsilon,$$

for all $t \geq 0$. The conclusion of Theorem B, in this case, again follows from Corollary 5.5 of [L2].

Remark 3. The argument of Theorem B can be simply modified for the case $d_j = \pm 1$ in [L3]. Suppose $(\hat{a}, \hat{d})$ is the nondegenerate saddle point or local maximum point of the renormalized energy W (cf. [BBH]), where $\hat{a} = (a_1, \ldots, a_{d+2\ell})$, $\hat{d} = (d_1, \ldots, d_{d+2\ell})$, $d_j = 1$ for $j \leq d + \ell; -1$ for $j > d + \ell$, and $\ell \in N$ is a constant. Then there is still a linking structure in $B_{\delta_0}(\hat{a})$, where $\delta_0 > 0$ is small. We also have the embedding (cf. Section 1), the gradient flow with intrinsic energy estimates (cf: [L], [L2]), and the associated projection lemma which is similar to Lemma 1. Hence we obtain the same result as Theorem B for this case. Note: The problem for general higher degree case $|d_j| > 1$ is still open.

Lecture 3. The Dynamical Law of Ginzburg-Landau Vortices

1. Gor'kov-Eliashberg Equation

Before we examine the dynamics of vortices, we have to look at the Gor'kov-Eliashberg evolution equation:

$$\left.\begin{aligned}
\eta\,\frac{\partial u}{\partial t} &+ i\eta\kappa\,\Phi\,u + \left(\tfrac{i\nabla}{\kappa} + A\right)^2 u - u + |u|^2\,u = 0 \\[2mm]
\frac{\partial A}{\partial t} &+ \nabla\Phi + \operatorname{curl}\operatorname{curl} A = -\frac{i}{2\kappa}\left(u^*\nabla u - u\nabla u^*\right) - A|u|^2,
\end{aligned}\right\} \tag{1.1}$$

in $\Omega \times R_+$, $\Omega \subset \mathbb{R}^3$.

In system (1.1), u is the complex order parameter, A is the magnetic (real) potential, κ, η are positive constants. Here u^* denotes the complex conjugate of u and Ω denotes a smooth bounded domain in $\mathbb{R}^3$, Φ is a scalar electric potential. The system (1.1) is supplemented by the initial and boundary conditions:

$$u(x,\,o) = u_o(x), \qquad A(x,\,o) = A_o(x), \qquad x \in \Omega; \tag{1.2}$$

$$\left.\begin{aligned}
\left(\frac{1}{\kappa}\nabla + A\right) u \cdot n &= 0, \\
\operatorname{curl} A \wedge \vec{n} &= 0, \quad (H_0 = 0), \\
\left(\frac{\partial A}{\partial t} + \nabla\Phi\right) \cdot \vec{n} &= \vec{E} \cdot \vec{n} = 0, \quad \text{on} \quad \partial\Omega.
\end{aligned}\right\} \tag{1.3}$$

Here $\vec{n}$ is the exterior unit normal along $\partial\Omega$.

Note that (1.1)–(1.3) is gauge-invariant, in the sense that if (u, A, Φ) is a solution, then so is $(u_\chi, A_\chi, \Phi_\chi)$, where

$$u_\chi = u\,e^{iK\chi}, \qquad A_\chi = A + \nabla\chi, \qquad \Phi_\chi = \Phi - \frac{\partial\chi}{\partial t}. \qquad \text{(cf. [D], [GE])}$$

The global existence, uniqueness (up to gauge transformations) of classical solutions of (1.1)–(1.3) have been studied by various authors, [D], [LR], [CHJ]. In [TW], the long time behavior, in particular, the existence of the global attractor is also investigated. Here we shall sketch the proof of the asymptotic stability result, which shows that, as $t \to \infty$, (1.1)–(1.3) has a unique asymptotic limit up to gauge transformations.

Next we shall consider the vortex motion for the following nonlinear equation:

$$\begin{cases}
\dfrac{\partial u}{\partial t} &= \Delta u + \dfrac{1}{\epsilon^2} u(1 - |u|^2) \quad \text{in} \quad \Omega \times R_+ \\
u(x,t) &= g(x) \quad \text{for} \quad x \in \partial\Omega, \qquad t > 0, \\
u(x, 0) &= u_0(x) \quad \text{for} \quad x \in \Omega.
\end{cases} \tag{1.4}$$

Here Ω is a two-dimensional, smooth bounded domain, ϵ is a positive parameter, $u : \Omega \times R_+ \longrightarrow \mathbb{R}^2$, $g : \partial\Omega \longrightarrow \mathbb{S}^1$. The system (1.4) can be viewed as a simplified model for (1.1) (cf. [N]). A system similar to (1.4) also appears in a canonical way when one expands a large class of second order dissipative systems about bifurcation points, [K], [BKP]. It serves, therefore, as one of the fundamental models in the study of the dynamics of non-equilibrium patterns [PZM], [RS].

The dynamics of the vortices in the limit $\epsilon \to 0$ can be considered within the framework of a general program initiated by J. Neu [N], and later extended, and improved by many others [RS], [PR], [E]. They formally used the method of matched asymptotic expansions to derive equations of motion for vortices. To leading order in $\left(\frac{1}{\log \epsilon} \right)$, the equations are of the form:

$$m_i \frac{d\, a_i\,(t)}{dt} = -\nabla\, a_i W_g(a), \qquad i = 1, 2, \ldots d. \tag{1.5}$$

The constants m_i are called mobilities of the vortices. One of the key facts they have derived is $m_i \simeq |\log \epsilon|$. In fact, it is derived in [N], [E] that $m_i \equiv \log \frac{1}{\epsilon}$, for all i. We want to give a rigorous proof of this dynamical law.

We shall not address here the hydrodynamical limit of the above dynamical law for an infinite number of vortices. Some discussions may be found in recent articles [CRS] and [E2].

2. Uniqueness of Asymptotic Limit

As in [D], one may choose the so-called zero electric potential gauge for system (1.1)–(1.3). This amounts to solving

$$\frac{\partial \chi}{\partial t} = \Phi, \quad \text{and} \quad \text{at } t = 0, \tag{2.1}$$

$$\Delta \chi = -\text{div } A \quad \text{on} \quad \Omega \quad \text{with} \quad \nabla \chi \cdot n = -A \cdot n \quad \text{on} \quad \partial \Omega. \tag{2.2}$$

Thus, in this gauge, $\Phi \equiv 0$, and the system (1.1) reduces to the gradient flow of the energy-functional:

$$E(u, A) = \frac{1}{2} \int_\Omega \left[\left| \left(\frac{i}{\kappa} \nabla + A \right) u \right|^2 + \frac{1}{2} \left(|u|^2 - 1 \right)^2 + |\text{curl } A|^2 \right] dx. \tag{2.3}$$

As shown in [D] and [TW], the flow

$$\frac{dv}{dt} = -\text{ grad } E(v), \qquad v(0) = v_0, \tag{2.4}$$

$v = (u, A)$ has a global classical solution. Our main result concerning (2.4) is the following

Theorem 1.

$$V_\infty = \lim_{t \to \infty} v(t) \quad exists.$$

To describe the idea, we start with the O.D.E.

$$\begin{cases} \dfrac{dx}{dt} & = & -\text{grad } f(x), \quad x \in \mathbb{R}^N \\ x(0) & = & x_0 \, . \end{cases} \tag{2.5}$$

We assume $f \in C^2(B_1)$, $\nabla f(0) = 0$ and x_0 is close to 0.
Case (i). If $A = \nabla^2 f(0)$ is positive definite, then $X(t) \to 0$ (at exponential rate) as $t \to +\infty$.

Proof. Calculate $\frac{d}{dt}|\dot{x}|^2 = -2\langle \nabla^2 f(x) \cdot \dot{x}, \dot{x} \rangle \le -2\lambda |\dot{x}|^2$. Here we shall assume $|x|(t) \le \delta_0$, and $(\nabla^2 f(x)) \ge \lambda I$, whenever $|x| \le \delta_0$. Thus $|\dot{x}(t)| \le |\dot{x}(0)|e^{-\lambda t}$, $\forall t > 0$, and $|x(t)| \le |x_0| + \frac{1}{\lambda}|\dot{x}(0)| = |x_0| + |\frac{1}{\lambda}\nabla f(x_0)|$. We shall always assume x_0 is so close to the origin that $|x_0| + \frac{1}{\lambda}|\nabla f(x_0)| \le \delta_0$. Then the assumption $|x(t)| \le \delta_0$ is true for all $t > 0$, and thus $x(t) \to 0$ at the exponential rate as $t \to +\infty$.
Case (ii). $\det(A) \ne 0$. Then one has that

$$-\frac{d}{dt}(f(x) - f(0))^{1/2} = \frac{1}{2}\frac{|\nabla f(x)||\dot{x}|}{(f(x) - f(0))^{1/2}}, \quad \text{if } f(x) > f(0) \tag{2.6}$$

and that

$$-\frac{d}{dt}(f(0) - f(x))^{1/2} = \frac{1}{2}\frac{|\nabla f(x)||\dot{x}|}{(f(0) - f(x))^{1/2}}, \quad \text{if } f(x) < f(0). \tag{2.7}$$

We obtain the following:
Either there is a $T \in (0, \infty)$ such that

$$f(x)(T) \le f(0) - \delta_0 \quad (\text{ for some } \delta_0 > 0)$$

Or

$$\lim_{t \to +\infty} x(t) = x_\infty \quad \text{exists} .$$

Indeed, if x is close to 0, then

$$\frac{|\nabla f(x)|}{|f(x) - f(0)|^{1/2}} \ge \frac{1}{2}\frac{\lambda_{\min}}{\lambda_{\max}} = C(A) > 0 .$$

Here $\lambda_{\min}$ and $\lambda_{\max}$ are the minimum and maximum, respectively, eigenvalues of $(A^T A)^{1/2}$. Therefore

$$\int_0^\infty |\dot{x}(t)| \, dt \le \frac{2\delta_0^{1/2}}{C(A)},$$

by (2.6), (2.7), whenever $f(x)(t) \ge f(0) - \delta_0$ for all $t \ge 0$.
    In such a case the conclusion $\lim_{t \to +\infty} x(t) = x_\infty$ follows (in fact $x_\infty = 0$) as for the case (i).

Case (iii).   det $(A) = 0$ and $f(x)$ is real analytic in B_1. We have the following well-known estimate (cf. [SL2]). There are two positive constants θ_0, $\sigma_0 \in (0, 1)$ depending on f such that

$$|f(x) - f(0)|^{\theta_0} \leq |\operatorname{grad} f(x)| \quad \text{whenever} \quad x \in B_{\sigma_0}(0), \tag{2.8}$$

$\nabla f(0) = 0$. Then, as for the case (ii), one has:
either there is a $T \in (0, \infty)$ such that

$$f(T) \leq f(0) - \delta_0 \quad \text{or} \quad \lim_{t \to \infty} x(t) = x_\infty$$

exists.

In [SL2], Simon considered the case where

$$E(u) = \int_M F(x, u, \nabla u)\, dx, \quad \text{and} \tag{2.9}$$

$$\begin{cases} \dot{u} &= -\operatorname{grad} E(u) \equiv \mathcal{M}(u) \\ u(0) &= u_0 \simeq 0, \quad \mathcal{M}(0) = 0. \end{cases} \tag{2.10}$$

where F is assumed to be analytic in both u and ∇u for u, ∇u near $\underline{O}$. Here M is a compact manifold without boundary.

Suppose L is elliptic, $Lu = \frac{d}{ds}\big|_{s=0} \mathcal{M}(sv)$. Then **either** there is a $T \in (0, \infty)$ such that

$$E(u(T)) \leq E(u(0)) - \delta_0, \quad \text{for some} \quad \delta_0 > 0,$$

or

$$u_\infty = \lim_{t \to \infty} u(t) \quad \text{exists}.$$

The key point is to show

$$|E(u) - E(0)|^{\theta_0} \leq \|\operatorname{grad} E(u)\|_{L^2} \tag{2.11}$$

for u near zero, and for some $\theta_0 \in (0, 1)$ (independent of u).

(2.11) plays the same role as (2.8). We can show (2.11) is valid for our functional $E(u, A)$ which is gauge invariant, degenerate (in a sense, not coercive). The conclusion of the theorem follows as for the case (iii) of (2.5). We refer to [DL] for more details.

3. Vortex Motion Equations

Let us consider the following model problems

$$\frac{1}{\lambda_\epsilon}\frac{\partial U_\epsilon}{\partial t} = \Delta U_\epsilon + \frac{1}{\epsilon^2} U_\epsilon (1 - |U_\epsilon|^2) \quad \text{in} \quad \Omega \times R_+, \tag{3.1}$$

$$U_\epsilon(x, 0) = U_\epsilon^0(x), \quad x \in \Omega \tag{3.2}$$

$$U_\epsilon(x, t) = g(x), \quad x \in \partial\Omega. \tag{3.3}$$

Here λ_ϵ is a constant depending only on $\epsilon > 0$. We shall always assume the following are true for the initial data in (3.2).

$$E_\epsilon(U_\epsilon^0) \leq \pi d \log \frac{1}{\epsilon} + K ; \qquad (H1)$$

$$U_\epsilon^0(x) \longrightarrow U_o(x) = \prod_{j=1}^{d} \frac{x - b_j}{|x - b_j|} \, e^{i\,h(x)} \qquad \text{weakly} \qquad (H2)$$

in $H_{\text{loc}}^1 \left(\overline{\Omega} \setminus \{b_1, \ldots, b_d\} \right)$ for some $h \in H^1(\Omega)$ and some d points $b_1, \ldots, b_d$ in Ω. Here $d = $ degree of $g : \partial\Omega \longrightarrow \mathbb{S}^2$, $d > 0$.

Our first result says vortices will not move (or move very slowly) in any finite time, when $\epsilon \to 0^+$, for solutions of (1.4) with initial data satisfying (H1)–(H2).

Theorem 2. *Assume* $\lambda_\epsilon \equiv 1$. *Then, as* $\epsilon \to 0$, *solutions* $U_\epsilon(x, t)$ *of (3.1)–(3.3) converge, for any* $t \geq 0$, *to* $U_0(x, t) = \prod_{j=1}^{d} \frac{x - b_j}{|x - b_j|} \, e^{i\,h(x, t)}$, *weakly in* $H_{\text{loc}}^1 \left(\overline{\Omega} \setminus \{b_1, \ldots, b_d\} \right)$. *Moreover, the function* h *satisfies*

$$\left. \begin{array}{rcl} \dfrac{\partial h}{\partial t} & = & \Delta h \quad in \quad \Omega \times R_+ \, , \\[2mm] h(x, 0) & = & h(x), \quad x \in \Omega \, , \\[2mm] \displaystyle\prod_{j=1}^{d} \dfrac{x - b_j}{|x - b_j|} \, e^{i\,h(x, t)} & = & g(x), \quad for \quad x \in \partial\Omega \, . \end{array} \right\} \qquad (3.4)$$

We shall not give the proof of this theorem (cf. [L]). Instead, we point out that one of the key points in the proof of the above theorem is to locate the vortices in the flow. We used a quantity of the form

$$\int_\Omega \rho(x) \left[|\nabla U_\epsilon|^2 + \frac{1}{2\epsilon^2} \left(|U_\epsilon|^2 - 1 \right)^2 \right] (x, t) \, dx , \qquad (3.5)$$

for some properly chosen $\rho(x)$, to achieve this purpose. The proof involves some crucial lower energy bounds, obtained in [BBH] and [St], and measurement of changes of the stated quantity (3.5). The proof can be generalized to obtain the following:

Corollary 1. *If* $\lambda_\epsilon \to \infty$, *and* $\lambda_\epsilon / \log \frac{1}{\epsilon} \to 0$, *as* $\epsilon \to 0$, *then the solutions,* $u_\epsilon(x, t)$ *of (3.1)–(3.3), converge to* $\prod_{j=1}^{d} \frac{x - b_j}{|x - b_j|} \, e^{i\,h_b(x)}$, *weakly in* $H_{\text{loc}}^1 \left(\overline{\Omega} \setminus \{b_1, \ldots, b_d\} \right)$, *where* $\Delta h_b = 0$ *in* Ω.

The corollary implies the mobilities of vortices are not less than the scale of $\log \frac{1}{\epsilon}$. Our next result shows mobilities cannot be larger than the scale $\log \frac{1}{\epsilon}$ either.

Theorem 3. *If $\lambda_\epsilon / \log \frac{1}{\epsilon} \to +\infty$ as $\epsilon \to 0$, then any possible weak limits of $U_\epsilon(\cdot, t)$, for any $t > 0$ are given by $\prod_{j=1}^{d} \frac{x - a_j}{|x - a_j|} e^{i h_a(x)}$. Here $\Delta h_a = 0$ in Ω, and $a = (a_1, \ldots, a_d)$ is a critical point of the renormalized energy $W_g(a)$ defined by: (cf. Lecture 2 and [BBH]) for $(x_1, \ldots, x_d) \in \Omega^d$,*

$$W_g(x_1, \ldots, x_d) = -\sum_{i \neq j} \log |x_i - x_j| - \sum_{j=1}^{d} R(x_j) \qquad (3.6)$$

$$+ \frac{1}{2\pi} \int_{\partial \Omega} \left(R + \sum_{j=1}^{d} \log |x - x_j| \right) g \wedge g_\tau ,$$

where

$$\begin{cases} \Delta R = 0 \quad in \quad \Omega \\ \dfrac{\partial R}{\partial n} = g \wedge g_\tau - \dfrac{\partial}{\partial \nu} \left(\sum_{j=1}^{d} \log |x - x_j| \right). \end{cases}$$

Finally we have the following dynamical law.

Theorem 4. *If $\lambda_\epsilon = \log \frac{1}{\epsilon}$, then, as $\epsilon \to 0$,*

$$U_\epsilon(x, t) \longrightarrow \prod_{j=1}^{d} \frac{x - a_j(t)}{|x - a_j(t)|} e^{i h_{a(t)}(x)} \quad weakly$$

$$in \quad H^1_{\text{loc}} \left(\Omega \setminus \{a_1(t), \ldots, a_d(t)\} \right), \quad for\ each \quad t \geq 0 .$$

Moreover, $\Delta h_{a(t)}(x) = 0$ in Ω, and

$$\begin{aligned} \dfrac{d\,a(t)}{dt} &= -\operatorname{grad} W_g(a(t)), \qquad 0 < t < \infty \\ a(0) &= b \equiv (b_1, \ldots, b_d) . \end{aligned} \left.\right\} \qquad (3.7)$$

Here $a(t) = (a_1(t), \ldots, a_d(t))$.

Thus we conclude that there are three time regimes. When $0 \leq t \leq \delta(\log \frac{1}{\epsilon})$, there is no vortices motion, but the phase function evolves with time according to the standard heat equation. For $\delta(\log \frac{1}{\epsilon}) \leq t \leq M \log \frac{1}{\epsilon}$, $\delta << 1$, $M >> 1$, the vortices motion obeys the law given by the O.D.E. in the statement of Theorem 4, and the phase function is a simple adjustment so that it will be harmonic in the domain and that it matches up the proper boundary conditions. When $t > M \log \frac{1}{\epsilon}$, both the location of vortices and the corresponding phase functions become static. Note that L. Simon's theorem also implies this (see Section 2 of this lecture) but for much larger time, say $t >> \frac{1}{\epsilon^2}$. Indeed, by using Theorem 3, Theorem 4 and the proof of Corollary 5.5 in [L], one can deduce the somewhat stronger statement that, in Theorem 3, as $\epsilon \to 0$, the limit of $U_\epsilon(x, t) \longrightarrow \prod_{j=1}^d \frac{x-a_j}{|x-a_j|} e^{i h_a(x)}$ is unique for all $t > 0$.

Now we sketch the proof of Theorem 4. It is in the same spirit as [L2], but with a somewhat different approach.

We consider a family of Radon measures

$$\mu_\epsilon(t) = \frac{1}{2} \left[|\nabla U_\epsilon|^2 (\cdot, t) + \frac{1}{2\epsilon^2} \left(|U_\epsilon|^2 (\cdot, t) - 1 \right)^2 \right] \frac{dx}{\pi \log \frac{1}{\epsilon}},$$

for each $t \geq 0$.

Proof of Theorem 4.

Step I. We show there is a sequence of $\epsilon_i \downarrow 0$, such that $\mu_{\epsilon_i}(t) \to \mu(t)$, as $i \to \infty$, for each $t \geq 0$. Here $\to$ denotes the weak convergence for measures.

To do so, we let $\phi \in C_0^1(\mathbb{R}^2)$, and calculate

$$
\begin{aligned}
\frac{d}{dt} \int_\Omega \phi^2(x) \, \mu_\epsilon(t) &= -\int_\Omega \phi^2 \left| \frac{\partial U_\epsilon}{\partial t} \right|^2 \cdot \frac{1}{\log^2 \frac{1}{\epsilon}} \, dx \\
&\quad -2 \int_\Omega \phi \nabla \phi \nabla U_\epsilon \frac{\partial U_\epsilon}{\partial t} \cdot \frac{1}{\log \frac{1}{\epsilon}} \, dx \\
&\leq \int_\Omega \phi^2 \cdot \mu_\epsilon(t) + C(\phi) \int_\Omega \left| \frac{\partial U_\epsilon}{\partial t} \right|^2 \cdot \frac{1}{\log \frac{1}{\epsilon}} \, dx \\
&\leq C(\phi) \big[\|\mu_\epsilon(0)\| + K'_\epsilon(t) \big]
\end{aligned}
$$

Here $C(\phi) = \|\phi\|_{C^1}^2$, $K_\epsilon(t) = \int_0^t \int_\Omega \left| \frac{\partial U_\epsilon}{\partial t} \right|^2 \frac{dx \, dt}{\log \frac{1}{\epsilon}}$ and $\|\mu_\epsilon(0)\|$ denotes the total measure of $\mu_\epsilon(0)$.

Thus the function $E_\epsilon(\phi, t) - C(\phi) [\|\mu_\epsilon(0)\| \cdot t + K_\epsilon(t)]$ is monotone non-increasing for $t \in [0, \infty)$. Here $E_\epsilon(\phi, t) = \int_\Omega \phi^2 \mu_\epsilon(t)$. We choose a countable dense subset $\{\phi_j\}_{j=1}^\infty$ of $C_0^1(\mathbb{R}^2)$, and for each j, we may find a sequence of ϵ's tending to zero such that the corresponding sequence of functions $\eta_\epsilon^j(t) \equiv E_\epsilon(\phi_j, t) - C(\phi_j) [\|\mu_\epsilon(0)\| \cdot t + K_\epsilon(t)]$ pointwise converges to a monotone non-increasing function $\eta_j(t)$, for each $t \geq 0$. Now we use the diagonal sequence to obtain a sequence of $\epsilon_n \downarrow 0$ such that $\eta_{\epsilon_n}^j(t)$ pointwise converges to a function $\eta(t)$, for each $t \geq 0$. It is then easy to see that $\mu_{\epsilon_n}(t) \to \mu(t)$ for each $t \geq 0$, and for a Radon measure $\mu(t)$. Here we also note that $\|\mu_\epsilon(0)\| \to d$ as $\epsilon \to 0^+$.

Step II. The Radon measure obtained in Step I is of the form: $\mu(t) = \sum_{j=1}^e \delta_{a_j(t)}$, for some d distinct points $a_1(t), \ldots, a_d(t)$ in Ω. Moreover, $\min\{|a_i(t) - a_j(t)|$, dist $(a_i(t), \partial\Omega), i \neq j, i, j = 1, \ldots, d\} \geq \delta_*$.

This is simply a consequence of the Compactness Theorem in Lecture 1. See also Theorem [BBH], part (iii).

Step III. The d-tuple point $a(t) = (a_1(t), \ldots, a_d(t))$ is continuously dependent on $t \in [0, \infty)$,

Indeed, one calculates

$$
\left| \frac{d}{dt} E_\epsilon(\phi, t) \right| \leq C(\phi) \int_\Omega \left| \frac{\partial U_\epsilon}{\partial t} \right|^2 \cdot \frac{1}{\log^2 \epsilon} \, dx +
$$

$$
+ C(\phi) \left[\int_\Omega |\nabla U_\epsilon|^2 \cdot \frac{1}{\log \frac{1}{\epsilon}} \, dx \right]^{1/2} \left[\int_\Omega \left| \frac{\partial U_\epsilon}{\partial t} \right|^2 \cdot \frac{1}{\log \frac{1}{\epsilon}} \, dx \right]^{1/2}
$$

$$
\leq C(\phi) \int_\Omega \left| \frac{\partial U_\epsilon}{\partial t} \right|^2 \cdot \frac{1}{\log^2 \epsilon} \, dx + C(\phi) \sqrt{\pi (d+1)} \ .
$$

$$
\cdot \left[\int_\Omega \left| \frac{\partial U_\epsilon}{\partial t} \right|^2 \cdot \frac{1}{\log \frac{1}{\epsilon}} \right]^{1/2}
$$

Using the fact that $\displaystyle \int_0^\infty \int_\Omega \left| \frac{\partial U_\epsilon}{\partial t} \right|^2 \cdot \frac{1}{\log \frac{1}{\epsilon}} \, dx \, dt \ \leq \ C(K, g, \Omega),$ and hence,

$$
\int_{t_1}^{t_2} \left[\int_\Omega \left| \frac{\partial U_\epsilon}{\partial t} \right|^2 \cdot \frac{1}{\log \frac{1}{\epsilon}} \, dx \right]^{1/2} dt \ \leq \ |t_2 - t_1|^{1/2} \cdot \sqrt{C(K, g, \Omega)},
$$

one can deduce the continuity of $a(t)$ by various proper choices of $\phi \in C_0^1(\Omega)$.

Step IV. Proof of (3.7). We first note it suffices to show (3.7) for $0 \leq t \leq \delta$, for some $\delta = \delta(g, \Omega, K) > 0$. In fact, we simply choose $\delta = \frac{\delta^*}{16}$. For $0 \leq t \leq \delta$, we have then $a_j(t) \subset B_\delta(a_j(0))$, $a_j(0) = b_j$, for $j = 1, \ldots, d$. For $j = 1, 2, \ldots, d$, $R \in [R_0/2, R_0]$, $R_0 = \frac{\delta^*}{4}$, we multiply (3.1), with $\lambda_\epsilon = \log \frac{1}{\epsilon}$, by ∇U_ϵ and use integration by parts in $B_R(b_j)$ to obtain

$$
\frac{1}{4\epsilon^2} \int_{\partial B_R(b_j)} (|U_\epsilon|^2 - 1)^2 \, \nu + \frac{1}{2} \int_{\partial B_R(b_j)} |\nabla U_\epsilon|^2 \nu - \int_{\partial B_R(b_j)} \frac{\partial U_\epsilon}{\partial \nu} \cdot \nabla U_\epsilon
$$

$$
= \frac{1}{\log \frac{1}{\epsilon}} \int_{B_R(b_j)} \frac{\partial U_\epsilon}{\partial t} \cdot \nabla U_\epsilon \, dx \ .
$$

On the other hand, one calculates, with $e_\epsilon = \frac{1}{2} \left[|\nabla U_\epsilon|^2 + \frac{1}{2\epsilon^2} (|U_\epsilon|^2 - 1)^2 \right]$, that

$$
\frac{1}{\log \frac{1}{\epsilon}} \frac{d}{dt} \int_{B_R(b_j)} x \cdot e_\epsilon(u) \, dx
$$

$$
= - \frac{1}{(\log \frac{1}{\epsilon})^2} \int_{B_R(b_j)} x \left| \frac{\partial U_\epsilon}{\partial t} \right|^2 - \frac{1}{\log \frac{1}{\epsilon}} \int_{B_R(b_j)} \frac{\partial U_\epsilon}{\partial t} \cdot \nabla U_\epsilon \, dx
$$

$$
+ \frac{1}{\log \frac{1}{\epsilon}} \int_{\partial B_R(b_j)} x \frac{\partial U_\epsilon}{\partial \nu} \cdot \frac{\partial U_\epsilon}{\partial t} \ .
$$

Therefore, by integrating with respect to $R \in \left[\frac{R_0}{2}, R_0\right]$, and with respect to the variable t, one has $f_\epsilon(t) = g_\epsilon(t) + h_\epsilon(t)$, where:

$$
f_\epsilon(t) = \fint_{R_0/2}^{R_0} \int_{B_R(b_j)} x \cdot \left[\frac{e_\epsilon(u)(t)}{\log \frac{1}{\epsilon}} - \frac{e_\epsilon(u)(0)}{\log \frac{1}{\epsilon}} \right] dx \, dR
$$

$$
g_\epsilon(t) = -\int_0^t \fint_{R_0/2}^{R_0} \left(\int_{\partial B_R(b_j)} G_\epsilon(u) \right) dR \, dt,
$$

$$
G_\epsilon(u) = \frac{(|U_\epsilon|^2 - 1)^2}{4\epsilon^2} \nu + \frac{1}{2} |\nabla U_\epsilon|^2 \nu - \frac{\partial U_\epsilon}{\partial \nu} \cdot \nabla U_\epsilon,
$$

$$
h_\epsilon(t) = \fint_{R_0/2}^{R_0} \int_0^t \frac{1}{(\log \frac{1}{\epsilon})^2} \int_{B_R(b_j)} x \left| \frac{\partial U_\epsilon}{\partial t} \right|^2 dx \, dt \, dR
$$
$$
+ \frac{2}{R_0 \log \frac{1}{\epsilon}} \int_0^t \int_{B_{R_0} \setminus B_{R_0/2}(b_j)} x \cdot \frac{\partial U_\epsilon}{\partial \nu} \cdot \frac{\partial U_\epsilon}{\partial t} \, dx \, dt
$$

Since $a_j(t) \in B_\delta(b_j)$, we have

$$
\int_{B_{R_0} \setminus B_{R_0/2}(b_j)} |G_\epsilon(u)| \, dx \leq \int_{B_{R_0} \setminus B_{R_0/2}(b_j)} \left[\frac{1}{4\epsilon^2} \left(|U_\epsilon|^2 - 1 \right)^2 + \frac{3}{2} |\nabla U_\epsilon|^2 \right] dx
$$
$$
\leq C(K, g, \Omega).
$$

Hence $g_\epsilon(t)$ is also lipschitz continuous on $[0, \delta]$ (independent of ϵ).

By **Step II**, $f_{\epsilon_n}(t) \longrightarrow \pi(a_j(t) - b_j)$ and hence $g_{\epsilon_n}(t) \longrightarrow \pi(a_j(t) - b_j)$. Finally we use a strong convergence theorem (cf. [L] Theorem 4.5) to conclude that

$$
\int_{B_{R_0} \setminus B_{R_0/2}(b_j)} G_{\epsilon_n}(u) \, dx \quad \text{converges to} \quad \frac{\pi}{2} R_0 \nabla_{a_j} W_g(a(t)).
$$ Therefore we obtain the identity:

$$
\pi \left(a_j(t) - a_j(0) \right) = -\pi \int_0^t \nabla_{a_j} W_g(a(t)) \, dt .
$$

This is exactly (3.7). Since the measure $\mu(t)$ obtained in Step I and Step II is completely identified by (3.7), we see $\mu_\epsilon(t) \longrightarrow \sum_{j=1}^d \delta_{a_j(t)}$ follows. $\qquad \square$

Remarks. (a) Recently, Jerrard and Soner [JS] proved the same dynamical law as (3.7) for the case that the initial data may have both $+1$ and -1 degree vortices. Of course (3.7) is then verified before the possible anhilation of vortices occurs. On the other hand, it is also possible that vortices of ± 1 degree may persist for all time, and the limiting static configurations may even be locally energy minimizing. (cf. [L3])

(b) Recently the author also derived rigorously the dynamic law for vortices under pinning effects. The idea can be generalized to the higher dimensions to study vortex-submanifold, in particular, filaments dynamics.

(c) It is a challenging open problem to study the dynamics of vortices for Schrödinger's equations:

$$
\begin{cases}
i\,\dfrac{\partial U_\epsilon}{\partial t} &= \Delta U_\epsilon + \dfrac{1}{\epsilon^2} U_\epsilon \left(1 - |U_\epsilon|^2\right), \quad \text{in } \Omega \times R\,, \\
U_\epsilon(x,\,0) &= U_\epsilon^0(x)\,, \qquad\qquad\quad x \in \Omega \\
U_\epsilon(x,\,t) &= g(x)\,, \qquad\qquad\qquad x \in \Omega\,.
\end{cases}
$$

The formal asymptotic analysis was carried out in [N] and [E].

There are also related questions in fluid dynamics. Similarly, one can also look at the corresponding hyperbolic equations. Many formal arguments there need to be justified.

(d) From the physics point of view, it will be very important to understand the dynamics of vortices under magnetic fields and applied currents. It is also important to study problems in the entire space and to study the stability of solutions.

References

[AB] L. Almeida and F. Bethuel, *Méthodes topologiques pour l'équation de Ginzburg-Landau*, C. R. Acad. Sci. Paris, **t. 320**, Série I, (1995), 935–939.

[AS] N. Andre and I. Shafrir, *Asymptotic behavior for the Ginzburg-Landau functional with weight, part I, II*, preprints.

[BBH] F. Bethuel, H. Brézis and F. Hélein, *Ginzburg-Landau vortices*, Birkhäuser, Boston, (1994).

[BCP] P. Bauman, N. Carlson and D. Phillips, *On the zeros of solutions to Ginzburg-Landau type systems*, SIAM J. Math. Anal. **24** (1993), 1283–1293.

[BKP] K. Bodenschats, W. Pesch and L. Kramer, *Structure and dynamics of dislocations in anisotropic pattern forming systems*, Phys. D **32** (1988), 135–145.

[BMR] H. Brézis, F. Merle and T. Rivière, *Quantization effects for* $-\Delta u = u(1 - |u|^2)$ *in* $\mathbb{R}^2$, Arch. Rat. Mech. Anal. **126** (1994), 35–58.

[BR] F. Bethuel and T. Rivière, *Vortices for a variational problem related to superconductivity*.

[CDG] S. Chapman, Q. Du and M. Gunzburger, *A variable thickness model for superconductivity thin films*, preprint.

[CET] Zhiming Chen, C.M. Elliot and Q. Tang, *Justification of a two-dimensional evolutionary Ginzburg-Landau superconductivity model*, preprint.

[CHJ] Z. M. Chen, K. H. Hoffmann and L. S. Jiang, *On the Lawrence-Doniach model for layered superconductors*, preprint.

[CHO] J. Chapman, S. Howison and J. Ockendon, *Mareoscopic models for superconductivity*, SIAM Review **34** (1992), 529–560.

[CK] S. Chanillo and M. Kiessling, *Symmetry of solutions of Ginzburg-Landau equations*, preprint.

[CRS] S. Chapman, J. Rubinstein and M. Schatzman, *A mean field model of superconducting vortices*, Euro. J. Appl. Math. (to appear).

[D] Q. Du, *Global existence and uniqueness of solutions of the time-dependent Ginzburg-Landau models for Superconductivity*, to appear in Applicable Analysis.

[De] De Gennes, *Superconductivity of metals and alloys*, Addison-Wesley Publishing Company.

[Di] S. Ding and Z. Liu, *Pinning of vortices for a variational problem related to the superconducting thin film having variable thickness*, preprint.

[DF] M. del Pino and P. L. Felmer, *Local minimizers for Ginzburg-Landau energy*, preprint (1995).

[DG] Q. Du and M. D. Gunzburger, *A model for superconducting thin films having variable thickness*, Phys. D (Nonlinear Phenomena) **69** (1993), 215–231.

[DGP] Q. Du, M. D. Gunzburger and J. Peterson, *Analysis and approximation of Ginzburg-Landau models for superconductivity*, SIAM Review **34** March (1992).

[DL] Q. Du and F. H. Lin, *Ginzburg-Landau vortices, pinning and hysterisis*, preprint.

[E] E. Weinan, *Dynamics of vortices in Ginzburg-Landau theories, with applications to superconductivity*, Phys. D **77** (1994), 383–404.

[E2] ____, *Dynamics of vortex liquid in Ginzburg-Landau theories, with applications to superconductivity*, Phys. Review B. vol. **50** #2 (1994), 1126–1135.

[F] H. Federer, *Geometric Measure Theory*, Springer-Verlag, Heidelberg-Berlin-New York, (1969).

[FP] P. C. Fife and L. A. Peletier, *On the location of defects in stationary solutions of the Ginzburg-Landau equations in $\mathbb{R}^2$*, Quart. Appl. Math. (to appear).

[GE] L. Gor'kov and G. Eliashberg, *Generalization of the Ginzburg-Landau equations for nonstationary problems in the case of alloys with paramagnetic impurities*, Soviet Phys. JETP **27** (1968), 328–334.

[GL] V. Ginzburg and L. Landau, *On the theory of superconductivity*, Zh. Eksper. Teoret. Fiz. **20** (1950), 1064–1082. [English translation in Men of Physics: L. Landau, I (D. ter. Haar ed.), Pergamon, New York and Oxford, (1965), 138–167.]

[GO] M. Guzburger and J. Ockendon, *Mathematical models in superconductivity*, SIAM News, November and December 1994.

[HL] R. Hardt and F. H. Lin, *Singularities for p-energy minimizing unit vector fields on planar domains*, Cal. of Variations and P.D.E. (to appear).

[JT] A. Jaffe and C. Taubes, *Vortices and Monopoles,* Birkhäuser, Boston and Basel, (1980).

[JS] R. Jerrard and M. Soner, *Dynamics of Ginzburg-Landau vortices*, preprint.

[K] Y. Kuramoto, *Chemical Waves, Oscillations, and Turbulence*, Springer-Verlag, (1984).

[LiL] E. H. Lieb and M. Loss, *Symmetry of Ginzburg-Landau minimizer in a disc*, Math. Res. Letters **1** (1994), 701–715.

[Lin] T. C. Lin, *Spectrum of linearized operators at the radial solutions of the Ginzburg-Landau equation*.

[LL] T. C. Lin and F. H. Lin, *Minimax solutions of Ginzburg-Landau equations*, preprint.

[L] F. H. Lin, *Solutions of Ginzburg-Landau equations and critical points of the renormalized energy*, Analyse non Linéaire, IHP. To appear.

[L2] ____, *Some dynamic properties of Ginzburg-Landau vortices; and A remark on the previous paper*, both to appear in CPAM.

[L3] ____, *Mixed vertex-antivertex solutions of Ginzburg-Landau equations*, Arch. Rat. Mech. Analysis (to appear).

[LR] C. Lefter and V. D. Radulescu, *On the Ginzburg-Landau energy with weight*, C. Rend. Acad. Sci. Paris. To appear.

[N]   J. Neu, *Vortices in complex scale fields*, Phys. D **43** (1990), 385–406.

[M]   P. Mironescu, *On the stability of radial solutions of the Ginzburg-Landau equations*, Jour. Functional Analysis, **130** (1995), 334–344.

[PeR] L. Peres and J. Rubinstein, *Vortex Dynamics for $U(1)$-Ginzburg-Landau Models*, Phys. D. **64** (1993), 299–309.

[PR]   L. Pismen and J. D. Rodriguez, *Mobilities of singularities in dissipative Ginzburg-Landau equations*, Phys. Rev. A **42** (1990), 2471–2474.

[PZM] Y. Pomeau, S. Zaleski and P. Manneville, *Disclinations in liquid crystals*, Quart. Appl. Math. **50** (1992), 535–545.

[RS]   J. Rubinstein and P. Sternberg, *On the slow motion of vortices in the Ginzburg-Landau heat flow*, preprint.

[S]   I. Shafrir, *Remarks on solutions of $Deu = u(1 - |u|^2)$ in $\mathbb{R}^2$*, C.R. Acad. Sci. Paris, **t. 318**, Serie I, (1994), 327–331.

[SL]   L. Simon, *Lectures on geometric measure theory*, Austr. Nat. Univ. CMA, (1983).

[SL₂] L. Simon, *Asymptotics for a class of non-linear equations with applications to geometric problems*, Annals of Math **118** (1983), 525–572.

[St]   M. Struwe, *On the asymptotic behavior of minimizers of the Ginzburg-Landau model in 2-dimensions*, J. Diff. Equations **7** (1994).

[TW]   Q. Tang and S. Wang, *Ginzburg-Landau equations of superconductivity*, preprint.

Progress in Nonlinear Differential Equations
and Their Applications, Vol. 29
© 1997 Birkhäuser Verlag Basel/Switzerland

Wave maps

Michael Struwe

Mathematik, ETH-Zentrum, CH-8092 Zürich

ABSTRACT. In these lectures we outline the known results concerning existence, uniqueness, and regularity for the Cauchy problem for harmonic maps from $(1 + m)$-dimensional Minkowski space into a Riemannian target manifold, also known as σ-models or wave maps. In particular, we mark the limits of the classical theory in high dimensions and trace recent developments in dimension $m = 2$, substantiating the conjecture that in this "conformal" case the Cauchy problem is well-posed in the energy space.

1. Local existence. Energy method

1.1. The setting

Let (M, γ) be an m-dimensional Riemannian manifold without boundary, the "domain" of our maps, and let (N, g) be a compact, k-dimensional Riemannian manifold, with $\partial N = \emptyset$, the "target". For simplicity, in these lectures we only consider the case $M = \mathbb{R}^m$; however, many of the results presented below can easily be extended to the case of a compact domain manifold, for instance, to the case $M = T^m = \mathbb{R}^m / \mathbb{Z}^m$, the flat torus. Moreover, by Nash's embedding theorem, we may assume that $N \subset \mathbb{R}^d$ isometrically for some $d > k$. We denote as $T_p N \subset T_p \mathbb{R}^d \cong \mathbb{R}^d$ the tangent space of N at a point p, and we denote as $T_p^\perp N$ the orthogonal complement of $T_p N$ with respect to the inner product $\langle \cdot, \cdot \rangle$ on $\mathbb{R}^d$. TN, $T^\perp N$ will denote, respectively, the corresponding tangent and normal bundles. Moreover, since N is compact, there exists a tubular neighborhood $U_{2\delta}(N)$ of width 2δ of N in $\mathbb{R}^d$ such that the nearest neighbor projection $\pi_N \colon U_{2\delta}(N) \to N$ is well-defined and smooth.

For M and N as above we consider smooth maps $u \colon \mathbb{R} \times M \to N \hookrightarrow \mathbb{R}^d$ on the space-time cylinder $\mathbb{R} \times M$. The space-time coordinates will be denoted as $z = (t, x) = (x^\alpha)_{0 \le \alpha \le m}$ and we denote as $\frac{\partial}{\partial x^\alpha} u = \partial_\alpha u = u_{x^\alpha}$ the partial derivative of u with respect to x^α, $0 \le \alpha \le m$. Also let $D = (\frac{\partial}{\partial t}, \nabla) = (\frac{\partial}{\partial x^\alpha})_{0 \le \alpha \le m}$. $\mathbb{R} \times M$ will be endowed with the Minkowski metric $\eta = (\eta_{\alpha\beta}) = \mathrm{diag}(1, -1, \ldots, -1)$ and we raise and lower indeces with the metric. By convention, we tacitly sum over repeated indeces. Thus, for example, $\partial^\alpha = \eta^{\alpha\beta} \partial_\beta$, where $(\eta^{\alpha\beta}) = (\eta_{\alpha\beta})^{-1} (= (\eta_{\alpha\beta})$ in our setting). Moreover,

$$\Box = \partial^\alpha \partial_\alpha = \frac{\partial^2}{\partial t^2} - \Delta$$

is the wave operator and

$$\frac{1}{2}\langle \partial^\alpha u, \partial_\alpha u\rangle = \frac{1}{2}\left(|u_t|^2 - |\nabla u|^2\right)$$

is the Lagrangean density of u.

1.2. Wave maps

Let $u: \mathbb{R} \times M \to N$ be sufficiently smooth. A (compactly supported) variation of u is a family of maps $u_\epsilon: \mathbb{R} \times M \to N$ depending smoothly on a parameter $\epsilon \in]-\epsilon_0, \epsilon_0[$ for some $\epsilon_0 > 0$, with $u_0 \equiv u$ and such that $u_\epsilon \equiv u_0$ outside some compact region $Q \subset \mathbb{R} \times M$ for all ϵ.

Given a map $\varphi \in C_0^\infty(\mathbb{R} \times M; \mathbb{R}^d)$, an admissible variation may be obtained, for instance, by letting $u_\epsilon = \pi_N(u + \epsilon\varphi)$ for $|\epsilon| < 2\delta \|\varphi\|_{L^\infty}^{-1}$, where $\pi_N: U_{2\delta}(N) \to N$ is the smooth nearest neighbor projection defined in Section 1.1.

A map u then is a wave map if u is a stationary point for the Lagrangean

$$\mathcal{L}(u; Q) = \frac{1}{2}\int_Q \langle \partial^\alpha u, \partial_\alpha u\rangle\, dz$$

with respect to compactly supported variations u_ϵ, $|\epsilon| < \epsilon_0$, in the sense that

$$\frac{d}{d\epsilon}\mathcal{L}(u_\epsilon; Q)\Big|_{\epsilon=0} = 0,$$

where Q strictly contains the support of $u_\epsilon - u$.

In particular, for the variation $u_\epsilon = \pi_N(u + \epsilon\varphi)$ above we then obtain the equation

$$0 = \frac{d}{d\epsilon}\mathcal{L}\big(\pi_N(u + \epsilon\varphi); Q\big)\Big|_{\epsilon=0} = \int_Q \langle \partial^\alpha u, \partial_\alpha(d\pi_N(u) \cdot \varphi)\rangle\, dz$$

$$= -\int_Q \langle \partial^\alpha \partial_\alpha u, d\pi_N(u) \cdot \varphi\rangle\, dz$$

for all $\varphi \in C_0^\infty(\mathbb{R} \times M; \mathbb{R}^d)$; that is, $\Box u(z) \perp T_{u(z)}N$ for all $z \in \mathbb{R} \times M$, or

$$\Box u \perp T_u N$$

for short.

To understand this relation in more explicit terms, fix a point $z_0 \in \mathbb{R} \times M$ and let $\nu_{k+1}, \ldots, \nu_d$ be an orthonormal frame for $T_p^\perp N$, smoothly depending on $p \in N$ for p near $p_0 = u(z_0)$. Then we can find scalar functions $\lambda^l: \mathbb{R} \times M \to \mathbb{R}$, $k < l \le d$, such that near $z = z_0$ there holds

$$\Box u = \lambda^l(\nu_l \circ u);$$

in fact,

$$\lambda^l = \langle \Box u, \nu_l \circ u\rangle$$
$$= \partial^\alpha \langle \partial_\alpha u, \nu_l \circ u\rangle - \langle \partial_\alpha u, \partial^\alpha(\nu_l \circ u)\rangle$$
$$= -\langle \partial_\alpha u, d\nu_l(u) \cdot \partial^\alpha u\rangle = -A^l(u)(\partial_\alpha u, \partial^\alpha u)$$

is given by the second fundamental form A^l of N with respect to ν_l. Thus, the wave map equation takes the form

$$\Box u = -A(u)(\partial_\alpha u, \partial^\alpha u) \perp T_u N, \tag{1.1}$$

where $A = A^l \nu_l$ is the second fundamental form of N. We regard the term on the right of (1.1) as a Lagrange multiplier associated with the target constraint $u(\mathbb{R} \times M) \subset N$.

1.3. Examples
In certain cases equation (1.1) takes a particularly simple form.

1.3.1.   The sphere.   For $N = S^k \subset \mathbb{R}^{k+1}$ equation (1.1) translates into the equation

$$\Box u = \big(|\nabla u|^2 - |u_t|^2 \big) u.$$

Indeed, since $u \perp T_u S^k$ it suffices to check that

$$\langle \Box u, u \rangle = \partial^\alpha \langle \partial_\alpha u, u \rangle - \langle \partial_\alpha u, \partial^\alpha u \rangle = |\nabla u|^2 - |u_t|^2.$$

1.3.2.   Geodesics.   Suppose $\gamma \colon \mathbb{R} \to N$ is a geodesic parametrized by arc-length and $u = \gamma \circ v$ for some map $v \colon \mathbb{R} \times M \to \mathbb{R}$. Compute

$$\Box u = \partial^\alpha \big(\gamma'(v) \partial_\alpha v \big) = \gamma''(v) \partial_\alpha v \partial^\alpha v + \gamma'(v) \Box v.$$

Note that γ' is parallel along γ; that is, $\gamma''(s) \perp T_{\gamma(s)} N$ for all $s \in \mathbb{R}$. Thus, u satisfies (1.1) if and only if v solves the linear, homogeneous wave equation $\Box v = 0$.

1.4. Basic questions
In view of the hyperbolic nature of equation (1.1), in particular, in view of Example 1.3.2, it is natural to study the Cauchy problem for equation (1.1) for (sufficiently) smooth initial data

$$(u, u_t) \,|_{t=0} = (u_0, u_1) \colon M \to TN. \tag{1.2}$$

The basic questions we shall ask are the following.

Local well-posedness in the smooth category: Does the initial value problem (1.1), (1.2) for smooth data always admit a unique smooth solution for small time $|t| < T$?

The smoothness hypothesis on the solution and the data may be weakened. In fact, for a function $u \in L^2_{\text{loc}}(\mathbb{R} \times M; N)$ it is possible to interpret equation (1.1) in the sense of distributions provided $Du \in L^2_{\text{loc}}(\mathbb{R} \times M)$.

More specifically, for $s \in \mathbb{N}_0$ we let $H^s(M; N) = \{ v \in H^{s,2}(M; \mathbb{R}^d); v(M) \subset N \}$ denote the Sobolev space of maps $v \colon M \to N$ such that v possesses square integrable distributional derivatives of any order up to s. Moreover, we say that $u \in L^2_{\text{loc}}(\mathbb{R} \times M; N)$ is a weak solution of (1.1), (1.2) of class H^s provided $(\frac{\partial}{\partial t})^\sigma u(t) \in$

$H^{s-\sigma}(M)$ for all $\sigma \leq s$, locally uniformly in t, and if u weakly solves (1.1) and assumes the initial data (1.2) in the sense of traces.

Then we can pose the problem of

Regularity: What is the minimal regularity of the data to ensure unique local solvability of (1.1), (1.2) in the same regularity class?

Global well-posedness: Does there exist a regularity class such that the Cauchy problem (1.1), (1.2) admits a unique solution in this class for all time?

We do not consider explicitly the issue of stability, that is, continuous dependence of solutions on the data. However, quite often stability is related to uniqueness.

1.5. Energy estimates

Let $e(u) = \frac{1}{2}|Du|^2$ be the energy density of a map $u\colon \mathbb{R} \times M \to N$, and let

$$E\big(u(t)\big) = \int_{\mathbb{R}^m} \big(e(u)\big)(t)\, dx$$

be the total energy of u at time t. Note that, if u solves (1.1), we have the conservation law

$$0 = \langle \Box u, u_t \rangle = \frac{d}{dt}\left(\frac{|u_t|^2}{2}\right) - \operatorname{div}\langle \nabla u, u_t \rangle + \langle \nabla u, \nabla u_t \rangle$$

$$= \frac{d}{dt} e(u) - \operatorname{div}\langle \nabla u, u_t \rangle.$$

Hence, if $Du(t)$ has compact support, upon integrating over $\mathbb{R}^m$ we find that

$$\frac{d}{dt} E\big(u(t)\big) = 0;$$

that is, total energy is conserved. A similar energy estimate also holds on light cones. In particular, it follows that the diameter of $\operatorname{supp}\big(Du(t)\big)$ grows with speed at most 1 and hence $Du(t)$ has compact support for all t whenever $\operatorname{supp}\big(Du(0)\big)$ is compact.

1.6. L^2-theory

The above energy inequality may be generalized to obtain a priori bounds for higher derivatives, as well. Consider the Cauchy problem

$$\Box u = f \quad \text{in} \quad \mathbb{R} \times M$$

$$u \,|_{t=0} = g, \; u_t \,|_{t=0} = h,$$

where f, g, and h are smooth functions such that $\operatorname{supp}(Du(0)) = \operatorname{supp}(h, \nabla g)$ is compact and $\operatorname{supp}(f(t))$ is compact for any t. Then we have

$$\frac{d}{dt} e(u) - \operatorname{div}(\nabla u u_t) = f u_t \leq |f||u_t|$$

and hence by Hölder's inequality

$$\|Du(t)\|_{L^2}\cdot\frac{d}{dt}\|Du(t)\|_{L^2} = \frac{d}{dt}E\big(u(t)\big) \leq \int_{\{t\}\times\mathbb{R}^m} |f|\,|u_t|\,dx$$

$$\leq \|f(t)\|_{L^2}\|u_t(t)\|_{L^2} \leq \|f(t)\|_{L^2}\|Du(t)\|_{L^2}.$$

It follows that

$$\frac{d}{dt}\|Du(t)\|_{L^2} \leq \|f(t)\|_{L^2}.$$

Similarly, for any multi-index $I = (i_0,\ldots,i_m) \in \mathbb{N}_0^{1+m}$ with $|I| = \Sigma_\alpha i_\alpha$, letting $\partial^I = \Pi_\alpha \partial_\alpha^{i_\alpha}$ we obtain

$$\frac{d}{dt}\|D\partial^I u(t)\|_{L^2} \leq \|\partial^I f(t)\|_{L^2} \tag{1.3}$$

for all t. Integrating in time, thus we find that for any $I \in \mathbb{N}_0^{1+m}$ there holds

$$\sup_{0\leq t\leq T} \|D\partial^I u(t)\|_{L^2} \leq \|D\partial^I u(0)\|_{L^2} + \int_0^T \|\partial^I f(t)\|_{L^2}\,dt.$$

and similarly for $T < 0$. Note that, using the equation $\Box u = f$, we can express any derivative $D\partial^I u(0)$ in terms of spatial derivatives of g and h of orders $|I|+1$ and $|I|$, and space-time derivatives of f at $t = 0$ of order $|I - 1|$, respectively. For instance,

$$u_{tt} = \Box u + \Delta u = f + \Delta g.$$

Letting

$$\|v\|_{L^{\infty,2}} = \sup_{0\leq t\leq T} \|v(t)\|_{L^2},$$

therefore for any $s \in \mathbb{N}_0$ we obtain the estimate

$$\|D^{s+1}u\|_{L^{\infty,2}} \leq T\|D^s f\|_{L^{\infty,2}} + C\big(\|D^{s-1}f(0)\|_{L^2} + \|\nabla^{s+1}g\|_{L^2} + \|\nabla^s h\|_{L^2}\big).$$

with a constant $C = C(s)$.

1.7. Local existence for smooth data

The results of the preceding section apply to obtain a-priori bounds for smooth solutions u to (1.1), (1.2) by letting $f := A(u)(\partial_\alpha u, \partial^\alpha u)$, etc.

The class of admissible data for $s \in \mathbb{N}_0$ is given by

$$H_c^{s+1}(M;TN) = \{(u_0, u_1) \in L_{\text{loc}}^2(M;TN);$$

$$u_0 \in H^{s+1}(M;\mathbb{R}^d), u_1 \in H^s(M;\mathbb{R}^d), \text{supp}(u_1, \nabla u_0) \subset\subset \mathbb{R}^m\}.$$

Note that by Sobolev's embedding $H^s \hookrightarrow L^\infty$ for $s > \frac{m}{2}$. Therefore, and using interpolation, whenever for some constant C_0 the estimate

$$\sup_{0\leq t\leq T} \|D^{s_0+1}u(t)\|_{L^2} \leq C_0 \tag{1.4}$$

is satisfied for some $s_0 > \frac{m}{2}$, then for any $s \in \mathbb{N}$ we can estimate

$$\|D^s\big(A(u)(\partial_\alpha u, \partial^\alpha u)\big)\|_{L^2} \leq C_s\big(1 + \|D^{s+1}u\|_{L^2}\big),$$

uniformly in $0 \leq t \leq T$. By (1.3), therefore

$$\frac{d}{dt}\|D^{s+1}u(t)\|_{L^2} \leq C_s\big(1 + \|D^{s+1}u(t)\|_{L^2}\big),$$

and Gronwall's inequality yields bounds for $D^{s+1}u$ in $L^{\infty,2}$, provided $(u_0, u_1) \in H_c^{s+1}(M; TN)$. In particular, the estimate (1.4) will be valid for some $T > 0$ if we fix a constant $C_0 > \|D^{s_0+1}u(0)\|_{L^2}$, a constant depending only on u_0, u_1, and s_0.

Similarly, by the contraction principle, one can show the existence of a unique solution u of class H^{s+1} on a small time interval $0 \leq t \leq T$, provided $(u_0, u_1) \in H_c^{s+1}(M; TN)$ for some $s > \frac{m}{2}$. In this way we obtain

Theorem 1.1 *Fix initial data $(u_0, u_1) \in H_c^{s_0+1}(M; TN)$, where $s_0 > \frac{m}{2}$. There exists $T > 0$ and a unique solution $u: [0, T] \times M \to N$ of (1.1), (1.2) such that*

$$\sup_{0 \leq t \leq T} \|D^{s_0+1}u(t)\|_{L^2} < \infty.$$

Moreover, if $(u_0, u_1) \in H_c^{s+1}(M; TN)$ for some $s > s_0$, then

$$\sup_{0 \leq t \leq T} \|D^{s+1}u(t)\|_{L^2} < \infty.$$

In particular, if u_0, u_1 are smooth, also the solution u is smooth on $[0, T] \times M$.

For more details and references, see for instance [27] or [31].

1.8. A slight improvement

The local existence theorem in the preceding section did not use the special structure of the nonlinear term in (1.1) nor its geometric interpretation.

Using the fact that the term on the right of (1.1) is a "null-form" in the derivatives of the components of $u = (u^1, \ldots, u^d)$, that is, the fact that

$$A(u)(\partial_\alpha u, \partial^\alpha u) = A_{ij}(u)\partial_\alpha u^i \partial^\alpha u^j$$

is a sum of bilinear forms whose symbol

$$A_{ij}(u)\xi_\alpha \eta^i \xi^\alpha \eta^j = A(u)(\eta, \eta)\xi_\alpha \xi^\alpha$$

vanishes on the null cone

$$\{\xi = (\xi_\alpha) \in \mathbb{R} \times \mathbb{R}^m; \xi_\alpha \xi^\alpha = 0\},$$

Klainerman-Machedon [32] obtained the following slight improvement of Theorem 1.1 in low dimensions.

Theorem 1.2 *Suppose* $m \leq 3$. *Then for any data* $(u_0, u_1) \in H_c^2(M; TN)$ *there exists a unique local solution* u *of class* H^2. *If* $(u_0, u_1) \in H^s, s > 2$, *then so is* u.

The proof in [32] uses special L^2-estimates for null forms and is quite involved. A very simple proof, based on energy estimates alone, however, can be given if one uses the *geometric* interpretation of (1.1). This observation is due to [49].

Proof. (i) First we derive local a-priori estimates for D^2u for smooth solutions u. Let ∂ be any first order derivative. Differentiating equation (1.1), we obtain

$$\Box(\partial u) = \partial \left(A(u)(\partial_\alpha u, \partial^\alpha u) \right) = dA(u)(\partial u, \partial_\alpha u, \partial^\alpha u) + 2A(u)(\partial_\alpha \partial u, \partial^\alpha u)$$

with data

$$(\partial u \mid_{t=0}, \partial u_t \mid_{t=0}) \in H_c^1(M; T\mathbb{R}^d).$$

Note that, by orthogonality $\langle u_t, A(u)(\cdot, \cdot) \rangle = 0$,

$$\langle \partial u_t, A(u)(\partial_\alpha \partial u, \partial^\alpha u) \rangle = -\langle u_t, dA(u)(\partial u, \partial_\alpha \partial u, \partial^\alpha u) \rangle.$$

Hence we obtain

$$\frac{d}{dt} E\left(\partial u(t) \right) = \int_{\{t\} \times M} \langle \Box(\partial u), \partial u_t \rangle \, dx$$

$$\leq C\|dA(u)\|_{L^\infty} \cdot \int_M |Du(t)|^3 |D^2u(t)| \, dx.$$

Since N is compact, dA is uniformly bounded on N. Moreover, by Sobolev's embedding, if $m = 3$ we can estimate

$$\int_M |Du(t)|^3 |D^2u(t)| \, dx \leq \|Du(t)\|_{L^6}^3 \|D^2u(t)\|_{L^2}$$

$$\leq C\|D^2u(t)\|_{L^2}^4 = CE\left(Du(t) \right)^2.$$

Similarly, if $m = 2$ we estimate

$$\int_M |Du(t)|^3 |D^2u(t)| \, dx \leq C\|Du(t)\|_{L^2} \|D^2u(t)\|_{L^2}^3$$

$$\leq CE\left(u(t) \right)^{1/2} E\left(Du(t) \right)^{3/2}.$$

Thus, in both cases we arrive at a Gronwall type inequality

$$\frac{d}{dt} E\left(Du(t) \right) \leq CE\left(Du(t) \right)^\gamma$$

with some $\gamma > 1$. A local-in-time H^2-bound follows.

(ii) To show uniqueness in the class H^2 let u and v be solutions of (1.1) on $[0, T] \times M$ sharing Cauchy data

$$(u \mid_{t=0} = v \mid_{t=0}, u_t \mid_{t=0} = v_t \mid_{t=0}) \in H_c^2(M; TN).$$

Observe that, since $\langle u_t, A(u)(\cdot, \cdot)\rangle = 0$, etc. we have

$$
\begin{aligned}
\langle u_t - v_t, \Box u - \Box v\rangle &= \langle u_t - v_t, A(u)(\partial_\alpha u, \partial^\alpha u) - A(v)(\partial_\alpha v, \partial^\alpha v)\rangle \\
&= \langle u_t, \big(A(u) - A(v)\big)(\partial_\alpha v, \partial^\alpha v)\rangle - \langle v_t, \big(A(u) - A(v)\big)(\partial_\alpha u, \partial^\alpha u)\rangle \\
&\leq C|u - v||D(u - v)|\big(|Du|^2 + |Dv|^2\big).
\end{aligned}
$$

Thus, if $m = 3$, the energy inequality and Sobolev's embedding give

$$
\begin{aligned}
\frac{d}{dt} E\big((u - v)(t)\big) &= \int_{\{t\}\times M} \langle u_t - v_t, \Box u - \Box v\rangle \, dx \\
&\leq C \int_{\{t\}\times M} |u - v||D(u - v)|\big(|Du|^2 + |Dv|^2\big) \, dx \\
&\leq C\big(\|Du(t)\|_{L^6}^2 + \|Dv(t)\|_{L^6}^2\big)\|(u - v)(t)\|_{L^6}\|D(u - v)(t)\|_{L^2} \\
&\leq C\big(\|D^2 u(t)\|_{L^2}^2 + \|D^2 v(t)\|_{L^2}^2\big)\|D(u - v)(t)\|_{L^2}^2 \\
&\leq C\big(\|D^2 u\|_{L^{\infty,2}}^2 + \|D^2 v\|_{L^{\infty,2}}^2\big)E\big((u - v)(t)\big),
\end{aligned}
$$

and uniqueness follows. For $m = 2$ the argument is similar.

(iii) An H^2-solution preserves higher regularity of the data. Indeed, by Theorem 1.1 it suffices to show this for data $(u_0, u_1) \in H_c^3(M; TN)$. Let $u: [0, T] \times M \to N$ be a local H^2-solution of (1.1), (1.2). We claim that u is of class H^3 on $[0, T]$ as well. For this, by Theorem 1.1 it suffices to prove an a-priori estimate for $\|D^3 u(t)\|_{L^2}$. As before let ∂ be a first order differential operator. For simplicity, at first we consider only spatial derivatives.

Note that

$$
\begin{aligned}
\langle \partial^2 u_t, \partial^2\big(A(u)(\partial_\alpha u, \partial^\alpha u)\big)\rangle &= 2\langle \partial^2 u_t, A(u)(\partial_\alpha \partial^2 u, \partial^\alpha u) + A(u)(\partial_\alpha \partial u, \partial^\alpha \partial u)\rangle \\
&\quad + 4\langle \partial^2 u_t, dA(u)(\partial u, \partial_\alpha \partial u, \partial^\alpha u)\rangle + \langle \partial^2 u_t, d^2 A(u)(\partial u, \partial u, \partial_\alpha u, \partial^\alpha u)\rangle \\
&= \mathrm{I} + \mathrm{II} + \mathrm{III}.
\end{aligned}
$$

Clearly, by Sobolev's embedding theorem we can bound

$$
\begin{aligned}
\int_{\{t\}\times M} |\mathrm{III}| \, dx &\leq C \int_{\{t\}\times M} |Du|^4 |D^3 u| \, dx \\
&\leq C\|Du(t)\|_{L^\infty}\|Du(t)\|_{L^6}^3\|D^3 u(t)\|_{L^2} \\
&\leq C\|D^2 u(t)\|_{L^2}^3\big(1 + \|D^3 u(t)\|_{L^2}^2\big) \leq C\big(1 + E\big(D^2 u(t)\big)\big),
\end{aligned}
$$

$$
\begin{aligned}
\int_{\{t\}\times M} |\mathrm{II}| \, dx &\leq C \int_{\{t\}\times M} |Du|^2 |D^2 u||D^3 u| \, dx \\
&\leq C\|Du(t)\|_{L^6}^2\|D^2 u(t)\|_{L^6}\|D^3 u(t)\|_{L^2} \\
&\leq C\|D^2 u(t)\|_{L^2}^2\|D^3 u(t)\|_{L^2}^2 \leq C E\big(D^2 u(t)\big).
\end{aligned}
$$

Before we can estimate the first term in a similar fashion we need to express I in a more convenient way. Using orthogonality $\langle u_t, A(u)(\cdot, \cdot)\rangle$ again, we have

$$
\begin{aligned}
\langle \partial^2 u_t, A(u)(\partial_\alpha \partial^2 u, \partial^\alpha u)\rangle &= \partial\langle \partial u_t, A(u)(\partial_\alpha \partial^2 u, \partial^\alpha u)\rangle - \langle \partial u_t, A(u)(\partial_\alpha \partial^3 u, \partial^\alpha u)\rangle \\
&\quad - \langle \partial u_t, A(u)(\partial_\alpha \partial^2 u, \partial^\alpha \partial u)\rangle - \langle \partial u_t, dA(u)(\partial u, \partial_\alpha \partial^2 u, \partial^\alpha u)\rangle \\
&= \partial\langle \partial u_t, A(u)(\partial_\alpha \partial^2 u, \partial^\alpha u)\rangle + \langle u_t, dA(u)(\partial u, \partial_\alpha \partial^3 u, \partial^\alpha u)\rangle \\
&\quad + \langle u_t, dA(u)(\partial u, \partial_\alpha \partial^2 u, \partial^\alpha \partial u)\rangle - \langle \partial u_t, dA(u)(\partial u, \partial_\alpha \partial^2 u, \partial^\alpha u)\rangle;
\end{aligned}
$$

moreover,

$$
\begin{aligned}
\langle u_t, dA(u)(\partial u, \partial_\alpha \partial^3 u, \partial^\alpha u)\rangle &= \partial\langle u_t, dA(u)(\partial u, \partial_\alpha \partial^2 u, \partial^\alpha u)\rangle \\
&\quad - \langle \partial u_t, dA(u)(\partial u, \partial_\alpha \partial^2 u, \partial^\alpha u)\rangle - \langle u_t, d^2 A(u)(\partial u, \partial u, \partial_\alpha \partial^2 u, \partial^\alpha u)\rangle \\
&\quad - \langle u_t, dA(u)(\partial^2 u, \partial_\alpha \partial^2 u, \partial^\alpha u)\rangle - \langle u_t, dA(u)(\partial u, \partial_\alpha \partial^2 u, \partial^\alpha \partial u)\rangle.
\end{aligned}
$$

Similarly,

$$
\begin{aligned}
\langle \partial^2 u_t, A(u)(\partial_\alpha \partial u, \partial^\alpha \partial u)\rangle &= \partial\langle \partial u_t, A(u)(\partial_\alpha \partial u, \partial^\alpha \partial u)\rangle \\
&\quad - \langle \partial u_t, dA(u)(\partial u, \partial_\alpha \partial u, \partial^\alpha \partial u)\rangle - 2\langle \partial u_t, A(u)(\partial_\alpha \partial^2 u, \partial^\alpha \partial u)\rangle,
\end{aligned}
$$

and

$$
\langle \partial u_t, A(u)(\partial_\alpha \partial^2 u, \partial^\alpha \partial u)\rangle = -\langle u_t, dA(u)(\partial u, \partial_\alpha \partial^2 u, \partial^\alpha \partial u)\rangle.
$$

Thus, as above we conclude that

$$
\begin{aligned}
\int_{\{t\}\times M} I\, dx &\le C \int_{\{t\}\times M} (|Du|^2|D^2 u| + |Du|^4)|D^3 u| + |Du||D^2 u|^3 \, dx \\
&\le C\big(1 + E\big(D^2 u(t)\big)\big).
\end{aligned}
$$

Hence there holds

$$
\frac{d}{dt} E\big(\nabla^2 u(t)\big) \le C\big(1 + E\big(D^2 u(t)\big)\big).
$$

For $\partial = \frac{\partial}{\partial t}$ we use the fact that $\partial^2 u_t = (\Box u)_t + \Delta u_t$. Repeating the previous computations, we can estimate

$$
\int_{\{t\}\times M} \langle \Delta u_t, \partial^2 A(u)(\partial_\alpha u, \partial^\alpha u)\rangle \, dx \le C\big(1 + E\big(D^2 u(t)\big)\big).
$$

Moreover, by Young's inequality

$$
\begin{aligned}
\langle \Box u_t, \partial^2\big(A(u)(\partial_\alpha u, \partial^\alpha u)\big)\rangle &= \langle \partial\big(A(u)(\partial_\alpha u, \partial^\alpha u)\big), \partial^2\big(A(u)(\partial_\alpha u, \partial^\alpha u)\big)\rangle \\
&\le C\big(|Du|^3 + |Du||D^2 u|\big)\big(|Du|^4 + |Du|^2|D^2 u| + |D^2 u|^2 + |Du||D^3 u|\big) \\
&\le C\big(|Du|^7 + |Du|^5|D^2 u| + |Du|^4|D^3 u| + |Du|^3|D^2 u|^2 \\
&\quad + |Du|^2|D^2 u||D^3 u| + |Du||D^2 u|^3\big) \\
&\le C\big(|Du|^7 + |Du|^4|D^3 u| + |Du|^2|D^2 u||D^3 u| + |Du||D^2 u|^3\big)
\end{aligned}
$$

and

$$\int_{\{t\}\times M} |Du|^7 \, dx \le C\|Du(t)\|_{L^\infty}\|Du(t)\|_{L^6}^6$$

$$\le C\|D^2u(t)\|_{L^2}^6\big(1+\|D^3u(t)\|_{L^2}\big) \le C\big(1+E\big(D^2u(t)\big)\big)^{1/2},$$

$$\int_{\{t\}\times M} |Du||D^2u|^3 \, dx \le C\|Du(t)\|_{L^6}\|D^2u(t)\|_{L^2}\|D^2u(t)\|_{L^6}^2$$

$$\le C\|D^2u(t)\|_{L^2}^2\|D^3u(t)\|_{L^2}^2 \le CE\big(D^2u(t)\big).$$

Thus we also find that

$$\frac{d}{dt}E\big(u_{tt}(t)\big) \le C\big(1+E\big(D^2u(t)\big)\big)$$

and hence that

$$\frac{d}{dt}E\big(D^2u(t)\big) \le C\big(1+E\big(D^2u(t)\big)\big),$$

which yields the desired a-priori bound.

(iv) Existence of local H^2-solutions can now be obtained as follows. Approximate the given data $(u_0, u_1) \in H_c^2(M; TN)$ by smooth data $(u_0^n, u_1^n) \in H_c^3(M; TN)$ in the topology of $H_c^2(M; TN)$. The approximate solutions u^n exist on a uniform time interval $0 \le t \le T$ and $\|D^2u^n(t)\|_{L^2}$ is uniformly bounded for $n \in \mathbb{N}$ and $0 \le t \le T$. Hence as $n \to \infty$ a sub-sequence

$$u^n \to u \qquad \text{weakly in } H^2\big([0, T] \times M\big)$$

and by Rellich's theorem $u^n \to u$ strongly in $H^1\big([0, T] \times M\big)$. In particular,

$$A(u^n)(\partial_\alpha u^n, \partial^\alpha u^n) \to A(u)(\partial_\alpha u, \partial^\alpha u) \quad \text{in } L^1,$$

and hence u solves (1.1) in the distribution sense. $\qquad\qquad\square$

It is conjectured that the initial value problem for (1.1), (1.2) is locally well-posed for data of class $H_c^{s+1}(M; TN)$, where $s \ge \frac{m-2}{2}$. In particular, in the "conformal case" $m = 2$, we expect the initial value problem to be locally well-posed for finite energy data, and hence, since energy is conserved (by classical solutions), we expect the existence of global unique solutions in this case. In Lecture 3 we will give some partial results in this regard.

1.9. Global existence, the case $m = 1$

If $m = 1$, as observed by Shatah [41] we can exchange the roles of x and t in our derivation of the conservation law

$$\frac{d}{dt}e(u) - \frac{d}{dx}\langle u_t, u_x \rangle = 0$$

in Section 1.5 to obtain

$$-\frac{d}{dx}e(u) + \frac{d}{dt}\langle u_x, u_t\rangle = \langle \Box u, u_x\rangle = 0.$$

Taking the t-derivative of the first and the x-derivative of the second equation and adding, we thus find that $e(u)$ solves the linear homogeneous wave equation

$$\Box e(u) = \left(\frac{\partial^2}{\partial t^2} - \frac{\partial^2}{\partial x^2}\right)e(u) = 0. \tag{1.5}$$

From this fact we easily deduce:

Theorem 1.3 *Suppose $m = 1$ and let $(u_0, u_1) \in H_c^s(M; TN)$, $s \geq 2$. Then (1.1), (1.2) admits a unique global solution u of class H^s.*

Proof. By Sobolev's embedding, $H^1 \hookrightarrow L^\infty$ for $m = 1$. Hence, if $(u_0, u_1) \in H_c^2(M; TN)$

$$e(u)\mid_{t=0} \in H_c^1, \quad \frac{d}{dt}e(u) \in L^2$$

and the energy inequality applied to (1.5) yields the a-priori bound

$$E\big(e(u)(t)\big) \leq E\big(e(u)(0)\big) < \infty,$$

uniformly in $t \in \mathbb{R}$. Hence, by Sobolev's embedding again, $e(u)$ is uniformly a-priori bounded on space-time in terms of the data. The assertion of the Theorem then follows by the same reasoning as used in the proof of Theorem 1.1. $\qquad\square$

2. Blow-up and non-uniqueness

2.1. Overview

In Lecture 1 we convinced ourselves that the initial value problem

$$\Box u = -A(u)(\partial_\alpha u, \partial^\alpha u) \perp T_u N \quad \text{on } \mathbb{R} \times M, \tag{2.1}$$

$$u\mid_{t=0} = u_0, u_t\mid_{t=0} = u_1 \quad \text{on } M, \tag{2.2}$$

is locally well-posed for sufficiently regular initial data $(u_0, u_1) \in H_c^{s+1}(M; TN)$, $s > \frac{m}{2}$, see Theorem 1.1. In this lecture we now investigate the behaviour of solutions for large time. In particular, depending on the dimension m of the domain and geometric properties of the target, we shall observe a decay of regularity, blow-up of higher derivatives and non-uniqueness of weak solutions beyond such blow-up points. These results indicate the limits of a regularity theory for (2.1) in dimensions $m \geq 3$ and raise the question whether there exists a class of weak solutions for which the initial value problem for wave maps is well-posed in a global sense. For better perspective and comparison, in the next section we give a brief survey of the known regularity results for harmonic maps in the elliptic (stationary) case and for the associated parabolic flow.

2.2. Regularity in the elliptic and parabolic cases

In the elliptic case we consider weak solutions $u \in H^1_{\text{loc}}(M; N)$ of the equation

$$- \Delta u = A(u)(\nabla u, \nabla u) \perp T_u N \tag{2.3}$$

with finite static energy

$$E_{st}(u) = \frac{1}{2} \int_M |\nabla u|^2 \, dx < \infty.$$

Here, M may be a smooth, compact m-manifold, possibly with boundary, or $M = \mathbb{R}^m$.

Associated with (2.3) is the heat flow

$$u_t - \Delta u = A(u)(\nabla u, \nabla u) \perp T_u N, \tag{2.4}$$

$$u\,|_{t=0} = u_0, \tag{2.5}$$

which is the L^2-gradient flow for E_{st} in the space $H^1(M; N)$.

2.2.1. Geometric constraints. If the sectional curvature K_N of N is non-positive, the Bochner identity for (2.3) implies that

$$-\Delta e(u) \leq 0$$

and hence an a-priori C^1-bound for smooth solutions. The same is true for the heat flow (2.4), (2.5). Hence the family $u(t, \cdot)$ of maps generated by (2.4), (2.5) is relatively compact in any C^l-topology and accumulates at a smooth limit $u_\infty \colon M \to N$.

Note, moreover, that (2.4) implies the identity

$$\frac{1}{2} \frac{d}{dt} |\nabla u|^2 - \operatorname{div}\langle \nabla u, u_t \rangle + |u_t|^2 = 0.$$

Upon integrating this equation over $[S, T] \times M$, we deduce that for any $T > S > 0$ there holds

$$\int_S^T \int_M |u_t|^2 \, dx \, dt + E_{st}\big(u(T)\big) \leq E_{st}\big(u(S)\big). \tag{2.6}$$

Letting $S \to 0$, $T \to \infty$, we find the a-priori estimate

$$\int_0^\infty \int_M |u_t|^2 \, dx \, dt \leq E_{st}(u_0), \ \sup_t E_{st}\big(u(t)\big) \leq E_{st}(u_0); \tag{2.7}$$

in particular, $u_t \to 0$ smoothly, as $t \to \infty$.

From this observation, Eells-Sampson [10] derived their fundamental existence result:

Theorem 2.1 *Suppose $K_N \leq 0$. Then for any smooth map $u_0 \colon M \to N$ there exists a smooth harmonic map $u_\infty \colon M \to N$ homotopic to u_0.*

In fact, for $K_N \leq 0$ *every* weakly harmonic map $u \in H^1(M; N)$ is smooth. The curvature constraint on the target can be replaced by the condition that the range of u is contained in a strictly convex geodesic ball on the target N or that the range $u(M)$ is the domain of a strictly convex function; see Hildebrandt [26] or Jost [28], [29] for a survey.

Theorem 2.1 may be false if the condition $K_N \leq 0$ is violated. The following result by Lemaire [33] and Wente [50] shows that, for instance, maps $u_0 \colon B^2 = B_1(0; \mathbb{R}^2) \to S^2$ of degree $\neq 0$ and which are constant on the boundary of B cannot be represented by harmonic maps.

Theorem 2.2 *If $u \in H^1(B^2; S^2)$ is harmonic with $u \mid_{\partial B} \equiv$ const., then $u \equiv$ const.*

Proof. Let $u \mid_{\partial B} \equiv p \in S^2$ and let $\pi \colon S^2 \setminus \{p\} \to \mathbb{R}^2$ denote stereographic projection from the antipodal point of p. Also let $i \colon \mathbb{R}^2 \to \mathbb{R}^2$ denote the involution $i(x) = -x$. Extend u to a map $\tilde{u} \in H^1(\mathbb{R}^2; S^2)$ by letting

$$\tilde{u}(x) = \pi^{-1}\left(i\big(\pi\big(u(\frac{x}{|x|^2})\big)\big)\right), \quad |x| > 1.$$

Since $\pi^{-1} \circ i \circ \pi$ induces an isometry of S^2 and since $\tilde{u}$ and $\nabla \tilde{u}$ are (weakly) continuous along ∂B, $\tilde{u}$ is weakly harmonic and hence smooth by Hélein's result, Theorem 2.4 below.

From (2.3) it follows by direct computation that the Hopf differential associated with $\tilde{u}$,

$$\Phi(x^1 + ix^2) = |\tilde{u}_{x^1}|^2 - |\tilde{u}_{x^2}|^2 - 2i\,\tilde{u}_{x^1} \cdot \tilde{u}_{x^2}$$

is a holomorphic function on $\mathbb{R}^2 \cong \mathbb{C}$.

Moreover, by conformal invariance of Dirichlet's integral, $\Phi \in L^1(\mathbb{R}^2)$, and hence, by the mean value property of holomorphic maps, $\Phi \equiv 0$. That is, $\tilde{u}$ is conformal. Since $\tilde{u} \equiv p$ on ∂B, we have $\nabla \tilde{u} \equiv 0$ on ∂B. But, by results of Hartman-Wintner [23], the branch points (where $\nabla \tilde{u} = 0$) of $\tilde{u}$ are isolated or $\tilde{u} \equiv$ const. $\square$

2.2.2. Partial Regularity. If we drop all geometric constraints on the target, for $m \geq 3$ there is no hope of proving regularity for weakly harmonic maps $u \in H^1(M; N)$. In fact, Rivière [37] constructed examples of weakly harmonic maps with finite energy which are everywhere discontinuous.

On the other hand, for maps u that minimize E among maps $v \in H^1(M; N)$ with the same boundary data, partial regularity results are available. Indeed, by results of Schoen-Uhlenbeck [39], [40] and Giaquinta-Giusti [18], energy-minimizing maps are smooth on the complement of a closed "singular set" of finite $(m - 3)$-dimensional Hausdorff measure. In particular, as was shown earlier by Morrey [36], energy-minimizing maps in dimension $m = 2$ are smooth. The example of the map $u(x) = \frac{x}{|x|} : M = B^m = B_1(0; \mathbb{R}^m) \to S^{m-1} = N$, which is energy minimizing if $m \geq 3$ [3], [34], shows that the above results cannot be improved.

Halfway between weakly harmonic maps and energy-minimizing maps lies the class of stationary maps $u \in H^1(M; N)$ which, by definition, are weakly harmonic and, in addition, are critical points of E_{st} also with respect to variations of the independent variables. By results of Evans [11] and Bethuel [2] these latter maps are smooth away from a singular set of dimension $\leq m - 2$.

There are analogous global existence and partial regularity results for the evolution problem (2.4), (2.5). In particular in the "conformal" case $m = 2$ we have the following result of Struwe [47], which was extended to the case $\partial M \neq \phi$ by Chang [4].

Theorem 2.3 *For any initial map $u_0 \in H^1(M; N)$ there exists a unique, global, weak solution $u : [0, \infty[\times M \to N$ of (2.4), (2.5), satisfying the energy inequality (2.6) for all $S < T$, and which is smooth away from finitely many points $(\bar{t}_i, \bar{x}_i), 1 \leq i \leq I \leq C E_{st}(u_0)$. Each singularity $(\bar{t}, \bar{x})$ is related to a smooth, non-constant harmonic map $\bar{u} : S^2 \to N$ in the sense that for suitable sequences $R_n \to 0, t_n \nearrow \bar{t}, x_n \to \bar{x}$ we have*

$$u_n(x) = u(t_n, \exp_{x_n}(R_n x)) \to \tilde{u} \quad in \ H^{2,2}_{loc}(\mathbb{R}^2; N)$$

as $n \to \infty$, where $\tilde{u} : \mathbb{R}^2 \to N$ is smooth, harmonic and extends to a smooth, non-constant harmonic map $\bar{u} : S^2 \to N$. (We refer to this "bubbling-off" process as "separation of spheres".)

Originally, uniqueness was only established among partially regular solutions as in the statement of Theorem 2.4. By results of Freire [13], [14] and Rivière [38], in case $N = S^k$ (and probably also for general targets) uniqueness also holds among weak solutions of class H^1 satisfying the energy inequality (2.6) for all $0 < S < T$.

By an example due to Chang-Ding-Ye [5], singularities actually may develop in finite time, even if the initial data are smooth. Theorem 2.3 therefore is best possible.

If $m \geq 3$, the existence of global, partially regular solutions to (2.4), (2.5) was derived in [6], based on the monotonicity formula and a-priori estimates for (2.4), (2.5) from [46]. However, there is no uniqueness in the energy class [9].

Also in the time-independent case the situation improves drastically if $m = 2$. In fact, we have the following beautiful result of Hélein [24].

Theorem 2.4 *Any weakly harmonic map $u \in H^1(M; N)$ is smooth.*

2.3. Regularity in the hyperbolic case

In short, one can say that all the problems with regularity of weakly harmonic maps and/or well-posedness of the evolution problem (2.4), (2.5) in the class of

H^1-solutions are present in the hyperbolic regime, as well. Thus, contrary to the title of this section, in the sequel we will not discuss any regularity properties of wave maps. Instead, we will show the break-down of regularity and loss of uniqueness for the initial value problem (2.1), (2.2) in dimensions $m \geq 3$. The examples we discuss indicate that there is hardly any regularity to be gained (in high dimension) from geometric conditions that we may impose on the target. Moreover, in order for (2.1), (2.2) to be well-posed in a suitable class, one still needs to identify a further "entropy condition" that will ensure uniqueness of weak solutions in this class. The situation in this regard thus is analogous to the situation for the parabolic evolution problem (2.4), (2.5) in dimensions $m \geq 3$.

With techniques available at this time we can therefore only hope to prove well-posedness of the initial value problem (2.1), (2.2) in case $m = 2$, analogous to Theorem 2.3 for the parabolic problem (2.4), (2.5) in this case. Some recent results in this regard will be presented in Lecture 3 .

2.3.1. Blow-up.
The simplest way to produce blow-up is to show the existence of self-similar solutions

$$u(t,x) = v(\frac{x}{|t|})$$

to (2.1) with non-constant smooth Cauchy data

$$u_0 = v, \quad u_1 = x \cdot \nabla v$$

at $t = -1$. Introduce similarity coordinates

$$\tau = \sqrt{t^2 - |x|^2}, \quad \xi = \frac{x}{|t|}$$

in the backward light cone from 0 and denote $|x| = r$, $|\xi| = \rho$, $x = r\omega$, $\xi = \rho\omega$ with $\omega \in S^{m-1}$. The Minkowski metric

$$ds^2 = dt^2 - dr^2 - r^2 d\omega^2$$

then can be expressed as

$$ds^2 = d\tau^2 - \tau^2 \left(\frac{d\rho^2}{(1-\rho^2)^2} + \frac{\rho^2}{1-\rho^2} d\omega^2 \right).$$

Hence u is stationary for the standard Lagrangean $\mathcal{L}$ if and only if $v(\xi) = v(\rho, \omega)$ is stationary for the reduced Lagrangean

$$\mathcal{L}_{sim}(v) = \frac{1}{2} \int \left\{ (1-\rho^2)^2 |v_\rho|^2 + \frac{1-\rho^2}{\rho^2} |v_\omega|^2 \right\} \frac{\rho^{m-1}}{(1-\rho^2)^{\frac{m+1}{2}}} \, d\rho \, d\omega$$

at $\tau = 1$. That is, u solves (2.1) if and only if v solves the equation

$$v_{\rho\rho} + \left(\frac{m-1}{\rho} + \frac{(m-3)\rho}{1-\rho^2} \right) v_\rho + \frac{1}{\rho^2(1-\rho^2)} \Delta_\omega v \perp T_v N. \qquad (2.8)$$

Remark that equation (2.8) is an *elliptic* harmonic map equation on the unit m-ball B with the hyperbolic metric

$$\frac{d\rho^2}{(1-\rho^2)^2} + \frac{\rho^2}{1-\rho^2}\, d\omega^2. \qquad (2.9)$$

We seek solutions v of equation (2.8) that extend smoothly to the "boundary" $\rho = 1$ of B and hence can be continued smoothly to all of $\mathbb{R}^m$. Since information propagates with speed ≤ 1 the unique solution u of (2.1), (2.2) with initial data

$$u_0 = v,\ u_1 = x \cdot \nabla v \quad \text{at } t = -1$$

then will coincide with $v(\frac{x}{|t|})$ inside the backward light cone $|x| \le -t$ and, if $v \not\equiv$ const. on B, we obtain blow-up at $t = 0$.

2.3.1.1. *The case $m = 2$.* If $m = 2$, equation (2.8) becomes

$$\left(\rho\sqrt{1-\rho^2}\, v_\rho\right)_\rho + \frac{1}{\rho\sqrt{1-\rho^2}}\Delta_\omega v \perp T_v N.$$

Multiplying by $\rho\sqrt{1-\rho^2}\, v_\rho$ and integrating with respect to $\omega \in S^1$, we obtain

$$\frac{d}{d\rho}\left(\int_{S^1} \rho^2(1-\rho^2)|v_\rho|^2\, d\omega - \int_{S^1} |v_\omega|^2\, d\omega\right) = 0,$$

and hence

$$\int_{S^1} \rho^2(1-\rho^2)|v_\rho|^2\, d\omega - \int_{S^1} |v_\omega|^2\, d\omega = C_0.$$

Inspection at $\rho = 0$ shows that $C_0 = 0$. Hence for $\rho = 1$ we obtain $v_\omega = 0$; that is, $v(1, \cdot) \equiv$ const.

Recall that by the Riemann mapping theorem the metric (2.9) on $B = B_1(0; \mathbb{R}^2)$ is locally conformal to the standard metric. In fact, define

$$\sigma(\rho) = \exp\left(-\int_\rho^1 \frac{d\rho}{\rho\sqrt{1-\rho^2}}\right)$$

and observe that the metric

$$d\sigma^2 + \sigma^2 d\omega^2 = \sigma^2\left(\frac{d\rho^2}{\rho^2(1-\rho^2)} + d\omega^2\right) = \left(\frac{\sigma}{\rho}\right)^2(1-\rho^2)\left(\frac{d\rho^2}{(1-\rho^2)^2} + \frac{\rho^2}{1-\rho^2}\, d\omega^2\right)$$

is conformal to the metric (2.9) on B.

That is, the map $(\rho, \omega) \mapsto (\sigma, \omega)$ is a conformal diffeomorphism ψ from B, endowed with the metric (2.9), to B with the standard metric.

By conformal invariance of Dirichlet's integral and hence of the harmonic map equation (2.3) in $m = 2$ dimensions, thus v induces a harmonic map $\bar{v} = v \circ \psi^{-1} \in H^1 \cap C^0(\bar{B}; N)$ on the standard disc with $\bar{v}|_{\partial B} \equiv$ const. By Lemaire's result Theorem 2.2, therefore we obtain the following result from [49].

Theorem 2.5 *If $m = 2$ and if $u(t, x) = v(\frac{x}{|t|})$ solves (2.1) for $|x| \le |t|$, where v extends to a smooth map on a neighborhood of $\overline{B_1(0)}$, then $v \equiv$ const.*

2.3.1.2 *The case $m \ge 3$.* In high dimensions, following Shatah-Tahvildar-Zadeh [45], we can obtain self-similar solutions to (2.1) as follows. We consider as target an m-dimensional manifold N with rotationally symmetric metric

$$ds^2 = dh^2 + g^2(h)\, d\omega^2$$

in spherical coordinates $h > 0$, $\omega \in S^{m-1}$ and we attempt to find solutions to (2.1) of the special form

$$u(t, r\omega) = h(t, r)\omega,$$

where we also express $x = r\omega \in \mathbb{R}^m$ in terms of spherical coordinates. Moreover, we make the ansatz $u(t, x) = v(\frac{x}{|t|})$, that is, $h(t, r) = \varphi(\frac{r}{|t|})$, $v(\xi) = \varphi(\rho)\omega$.

The reduced Lagrangean then becomes

$$\mathcal{L}_{sim}(v) = \frac{1}{2} \int \{(1 - \rho^2)^2 |\varphi_\rho|^2 + \frac{1 - \rho^2}{\rho^2}(m-1)g^2(\varphi)\} \frac{\rho^{m-1}}{(1 - \rho^2)^{\frac{m+1}{2}}}\, d\rho\, d\omega,$$

and equation (2.8) takes the form

$$\varphi_{\rho\rho} + \left(\frac{m-1}{\rho} + \frac{(m-3)\rho}{1 - \rho^2}\right)\varphi_\rho + \frac{(m-1)f(\varphi)}{\rho^2(1 - \rho^2)} = 0. \qquad (2.10)$$

where

$$f(\varphi) = g(\varphi)g'(\varphi).$$

For special target metrics g, equation (2.10) admits non-constant solutions φ for $0 < \rho \le 1$ that extend smoothly to all of $\mathbb{R}_+$. In fact, we may take

$$g^2(\varphi) = \varphi^2 - \frac{1}{2}\varphi^4 \qquad \text{for } 0 < \varphi < \varphi_0$$

for some fixed number $\varphi_0 > 0$ such that $1 < \varphi_0^2 < 2$, and we extend g smoothly to $\mathbb{R}_+$. If follows that

$$f(\varphi) = g(\varphi)g'(\varphi) = \frac{1}{2}\left(g^2(\varphi)\right)' = \varphi - \varphi^3$$

for $0 < \varphi < \varphi_0$. The linear function

$$\varphi(\rho) = c\rho,$$

where $c = \sqrt{\frac{2}{m-1}}$ then solves (2.10) for $0 < \rho < c^{-1}\varphi_0 = \sqrt{\frac{m-1}{2}}\varphi_0 = \rho_0$, and $\rho_0 > 1$ if $m \ge 3$. Note that for g as above, the radius of convexity of N around 0 is

$$\varphi_* = 1,$$

which is larger than c for $m \geq 4$ and equals c if $m = 3$. By changing the metric $g(\varphi)$ on N suitably for $\varphi > c$, and by changing the initial data for h off $\overline{B_1(0)}$, we may thus construct solutions to (2.10) which blow up in finite time, with initial data having compact support and such that the target manifold is convex, if $m \geq 4$, and only slightly fails to be convex, if $m = 3$.

A more detailed analysis shows that in 3 space dimensions blow-up may occur also for more general metrics on the target surface:

Theorem 2.6 (Shatah-Tahvildar-Zadeh [45]) *Suppose $g \in C^\infty$ satisfies $g(0) = 0$, $g'(0) = 1$ and suppose g' has a smallest positive zero φ_*. Also suppose that $g''(\varphi_*) \neq 0$. Then there is a class of regular initial data such that the corresponding Cauchy problem for equivariant harmonic maps from M^{3+1} into N has a solution that blows up in finite time.*

2.3.2. Non-uniqueness of weak solutions. In particular, Theorem 2.6 applies to the sphere, where $g(\varphi) = \sin \varphi$, $\varphi_* = \frac{\pi}{2}$. Shatah-Tahvildar-Zadeh construct a solution φ to (2.9) on $[0, \infty[$, satisfying

$$\varphi(1) = \varphi_*$$

and having the asymptotic expansion for $\rho \to \infty$:

$$\varphi(\rho) = a + \frac{b}{\rho} + \frac{d}{\rho^2} + O\left(\frac{1}{\rho^3}\right)$$

$$\varphi'(\rho) = -\frac{b}{\rho^2} + O\left(\frac{1}{\rho^3}\right).$$

They consider the corresponding function $h(t, r)\omega = \varphi\left(\frac{r}{t}\right)\omega$ as a weak solution of (2.1), that is,

$$h_{tt} - h_{rr} - \frac{2}{r}h_r + \frac{\sin 2h}{2r^2} = 0, \tag{2.11}$$

with singular initial data at $t = 0$, given by

$$h(0, r) = h_0(r) = a = \lim_{t \searrow 0} \varphi\left(\frac{r}{t}\right) \qquad (r \neq 0),$$

$$h_t(0, r) = h_1(r) = \frac{b}{r} = \lim_{t \searrow 0} \frac{d}{dt}\varphi\left(\frac{r}{t}\right) \qquad (r \neq 0). \tag{2.12}$$

Thereby, h is a weak solution of (2.11) say, on $[0, 1] \times \mathbb{R}^3$, if there holds

$$\int_0^1 \int_0^\infty \left\{-h_t\psi_t + h_r\psi_r + \frac{1}{2r^2}\psi \sin 2h\right\} r^2 \, dr \, dt = \int_0^\infty \psi(0, r)\frac{b}{r}r^2 \, dr \tag{2.13}$$

for any $\psi \in C^\infty\left([0,1] \times \mathbb{R}^3\right)$ such that $\psi(t, x) = \psi(t, r)$, $\psi(1, \cdot) \equiv 0$, and $\operatorname{supp}\psi(t) \subset B_R(0)$ for some $R > 0$. Moreover, h assumes the initial data (2.12) in the sense that

$$\|h(t,r) - a\|_{H^{1,2}_{\text{loc}}(\mathbb{R}^3)} \to 0 \qquad (t \to 0),$$

$$\left\|h_t(t,r) - \frac{b}{r}\right\|_{L^2(\mathbb{R}^3)} \to 0 \qquad (t \to 0).$$

Note that $h_0 \in H^{1,2}_{\text{loc}}$, $h_1 \in L^2_{\text{loc}}$.

On the other hand, also the function

$$\tilde{h}(t,r) = \begin{cases} \varphi\left(\frac{r}{t}\right), & r > t \\ \varphi_*, & r \leq t \end{cases}$$

weakly satisfies (2.11), (2.12) on $[0,1] \times \mathbb{R}^3$, with $D\tilde{h} \in L^\infty\left([0,1]; L^2(B_R(0))\right)$ for any $R > 0$, showing that weak solutions are in general not unique. To verify that $\tilde{h}$ solves (2.13), for any ψ we split

$$\int_0^1 \int_0^\infty \left\{-h_t\psi_t + h_r\psi_r + \frac{1}{2r^2}\psi \sin 2h\right\} r^2 \, dr \, dt - \int_0^\infty \psi(0,r)\frac{b}{r}r^2 \, dr$$

$$= \left\{\int_0^1 \int_t^\infty \{\dots\} r^2 \, dr \, dt - \int_0^\infty \psi(0,r)\frac{b}{r}r^2 \, dr\right\} + \int_0^1 \int_0^t \{\dots\} r^2 \, dr \, dt = I + II.$$

Clearly, since $D\tilde{h}(t,r) \equiv 0$ for $r \leq t$, the second integral $II = 0$. Moreover, since $\tilde{h} \equiv h$ for $r \geq t$, and since h satisfies (2.13) the first integral reduces to the boundary term

$$I = \frac{1}{\sqrt{2}} \int_0^1 (h_t(t,t) + h_r(t,t))\, \psi(t,t)t^2 \, dt$$

which also vanishes on account of

$$h_t + h_r = -\frac{r}{t^2}\varphi'\left(\frac{r}{t}\right) + \frac{1}{t}\varphi'\left(\frac{r}{t}\right)$$

$$= \frac{1}{t}\left(1 - \frac{r}{t}\right)\varphi'\left(\frac{r}{t}\right) = 0 \qquad \text{for } r = t.$$

Observe that $\tilde{h}$ induces a solution $\tilde{u}$ of (2.1) with $E(\tilde{u}(t); B_1(0)) > E(u(t); B_1(0))$ for any $t \in]0,1]$, where u is the solution corresponding to h. Hence there may be a chance of restoring uniqueness by some entropy principle.

3.   The conformal case $m = 2$

3.1.   Overview

The results presented in Lecture 2 leave little hope for the development of a satisfactory theory of existence, uniqueness, and stability for wave maps in high dimensions $m \geq 3$, even under very stringent geometric conditions on the target and/or very restrictive symmetry assumptions on the maps involved.

By contrast, as is illustrated by the absence of self-similar solutions, Theorem 2.5, the situation seems to be much better in dimension $m = 2$, due to conformal invariance of Dirichlet's integral in this dimension. Thus, we are tempted to conjecture that a result analogous to Theorem 2.3 for the "heat flow" related to harmonic maps of surfaces M also holds for the Cauchy problem for wave maps $u: \mathbb{R} \times M \to N$.

In this last of three lectures we will show that this conjecture is true for equivariant maps to surfaces of revolution and we sketch some recent developments towards a general theorem of well-posedness of the Cauchy problem in dimension 2.

3.2. The equivariant case
The results that follow are mostly due to Shatah-Tahvildar-Zadeh [44], [45] and Shatah-Struwe [42], [43].

3.2.1. Co-rotational maps. As in Lecture 2, Section 2.3, again we consider as target a surface of revolution N with metric

$$ds^2 = dh^2 + g^2(h) \, d\omega^2$$

written in terms of polar coordinates $h > 0$, $\omega \in S^1$. We assume that g is smooth with $g(-h) = -g(h)$ and $g'(0) = 1$. Moreover, we either suppose that

$$g(h) > 0 \qquad \text{if } h > 0 \tag{3.1}$$

and

$$\int_0^h |g(s)| \, ds \to \infty \qquad \text{as } h \to \infty, \tag{3.2}$$

or that there exists $q_1 > 0$ such that

$$g(q_1) = 0, \quad g(h) > 0 \qquad \text{for } 0 < h < q_1 \tag{3.3}$$

and g is odd around q_1 as well as around $q_0 = 0$ (and hence periodic of period $2q_1$). In the latter case, for ease of exposition only, in the following we will also assume that g is even around $\frac{q_1}{2}$.

Case (3.1), (3.2) corresponds to a non-compact target surface N, including the standard plane $g(h) \equiv h$ or metrics of negative curvature like $g(h) = sinh(h)$; condition (3.2) rules out sharp cusps "at infinity". Case (3.3) corresponds to a compact target, including the standard sphere $g(h) = \sin(h)$. Remark that (3.3) also implies (3.2).

Moreover, we consider maps $u: \mathbb{R} \times \mathbb{R}^2 \to N$ such that, expressing a point $x \in \mathbb{R}^2$ in polar coordinates $x = r\omega$, the angle $\omega \in S^1$ is preserved by u and $h(t, x) = h(t, r)$. Such maps will be called co-rotational.

For such u we have

$$\frac{1}{2\pi} \mathcal{L}(u) = \frac{1}{2} \int \left\{ |h_t|^2 - |h_r|^2 - \frac{g^2(h)}{r^2} \right\} r \, dr \, d\omega \, dt$$

and u is stationary for $\mathcal{L}$ if and only if $h: \mathbb{R} \times [0, \infty[\to \mathbb{R}$ satisfies

$$h_{tt} - h_{rr} - \frac{1}{r} h_r + \frac{f(h)}{r^2} = 0, \tag{3.4}$$

where
$$f(h) = g(h)g'(h).$$

Moreover, for smooth solutions the energy (scaled with a factor 2π)

$$\frac{1}{2\pi} E(u(t)) = \frac{1}{2} \int_0^\infty \left\{ |h_t|^2 + |h_r|^2 + \frac{g^2(h)}{r^2} \right\} r \, dr = E_{\text{equi}}(h(t))$$

is conserved.

Lemma 3.1 *If u is a co-rotational map with $E(u(t)) \leq$ const. for all $t \in \mathbb{R}$, then the associated map h is continuous on $\mathbb{R} \times \,]0, \infty[$ and $h(t, \cdot)$ extends continuously to $[0, \infty[$ for every $t \in \mathbb{R}$, where $g(h(t, 0)) = 0$ for all t.*

Proof. Since for any $r_0 > 0$ the integral

$$\int_{r_0}^\infty |h_r(t, r)|^2 \, dr \leq 2r_0^{-1} E_{\text{equi}}(h(t))$$

is uniformly bounded, by Sobolev's embedding $H^{1,2} \hookrightarrow C^{1/2}$ we conclude that $h(t, \cdot)$ is locally Hölder continuous on $]0, \infty[$, uniformly in $t \in \mathbb{R}$. Since $h_t \in L^2_{\text{loc}}(\mathbb{R} \times]0, \infty[)$, moreover, $h(\cdot, r)$ is continuous in t for almost every $r > 0$. Hence, the map $t \mapsto h(t, \cdot) \in C^0(]0, \infty[)$ is continuous by the theorem of Arzéla-Ascoli.

In order to prove continuity at $r = 0$, let

$$G(h) = \int_0^h |g(s)| \, ds.$$

Then, by Hölder's inequality, for any $t \in \mathbb{R}$ we have

$$\int_0^{r_0} |G(h)_r| \, dr \leq \int_0^{r_0} |g(h)||h_r| \, dr \leq \left(\int_0^{r_0} \frac{|g(h)|^2}{r} \, dr \right)^{1/2} \left(\int_0^{r_0} |h_r|^2 r \, dr \right)^{1/2}$$

$$\leq E_{\text{equi}}(h(t); B_{r_0}(0)) \to 0 \quad (r_0 \to 0).$$

$$(3.5)$$

It follows that $\lim_{r \to 0} G(h(t, r))$ exists for any t and hence, by strict monotonicity of G, that $\lim_{r \to 0} h(t, r) = h(t, 0)$ exists for any t. Finally, since

$$\int_0^\infty \frac{g^2(h)}{r} \, dr < \infty,$$

it follows that $g(h(t, 0)) = 0$ for any t. $\square$

In case (3.1) Lemma 3.1 implies the boundary condition

$$h(t, 0) = 0 \qquad \text{for all } t. \qquad (3.6)$$

In case of assumption (3.3), from Lemma 3.1 we only deduce that $h(t, 0) = dq_1$ for some $d \in \mathbb{Z}$.

Lemma 3.2 *There is a constant $\epsilon_0 > 0$ with the following property. If there exists $r_0 > 0$ such that*

$$E\big(u(t); B_{r_0}(0)\big) < \epsilon_0 \qquad \text{for all } t, \tag{3.7}$$

then $h(t,0)$ is constant in time.

Proof. From (3.5) we deduce that

$$\big|G\big(h(t,0)\big) - G\big(h(t,r_0)\big)\big| \le \epsilon_0$$

for all t. Since $h(t,r_0)$ depends continuously on t and since $G\big(h(t,0)\big) = d\,G(q_1)$ for some $d = d(t) \in \mathbb{Z}$, if $C\epsilon_0 < G(q_1)$ it follows that $h(t,0)$ is constant in time. $\square$

Finiteness of $E\big(u(t)\big)$ and finite propagation speed for (3.4) also implies the asymptotic boundary condition

$$h(t,r) \equiv q_0 \in \mathbb{R} \qquad \text{for } r \ge R_0 + |t| \tag{3.8}$$

for some number $R_0 > 0$ and some $q_0 \in \mathbb{R}$ such that $g(q_0) = 0$. We may normalize $q_0 = 0$. Together, (3.5) and (3.8) imply the uniform bound

$$\sup_r G\big(h(t,r)\big) \le \lim_{r_0 \to \infty} G\big(h(t,r_0)\big) + 2E_{\text{equi}}\big(h(t)\big) = G(0) + E_{\text{equi}}\big(h(t)\big).$$

From assumption (3.2) we then obtain

Lemma 3.3 *If $h\colon \mathbb{R} \times [0,\infty[\to \mathbb{R}$ corresponds to a co-rotational wave map u with data $(u_0, u_1) \in H_c^1$, then h is uniformly bounded.*

For later reference we also introduce the set

$$H_c^1\big([0,\infty[\big) = \Big\{(h_0, h_1); h_0 \in H_{\text{loc}}^{1,2} \cap C^0\big([0,\infty[\big), h_0(0) = d_0 q_1 \quad \text{for some } d_0 \in \mathbb{Z},$$

$$h_0(r), h_1(r) \equiv 0 \quad \text{for large } r\Big\}.$$

The initial value problem for co-rotational wave maps u with data $(u_0, u_1) \in H_c^1(\mathbb{R}^2; TN)$ at $t = 0$ then corresponds to the initial-boundary value problem (3.4) for data $\big(h\,|_{t=0}, h_t\,|_{t=0}\big) \in H_c^1\big([0,\infty[\big)$.

Due to energy conservation and the semi-linear character of (3.4) it is not hard to show the existence of a global weak solution to the initial-boundary value problem (3.4) of class H_c^1. However, it is not clear whether this solution is unique and whether the solution preserves any additional regularity properties of the data.

Fortunately, we can transform equation (3.4) to a form where these questions can be answered.

3.2.2.   The transformed equation.   To eliminate the singularity in (3.4), instead of h we introduce the map

$$\varphi(t,r) = \frac{h(t,r) - h(t,0)}{r}.$$

Recall that, by Lemma 3.2, $h(t,0) = dq_1$, where $d \in \mathbb{Z}$ is independent of $t \in \mathbb{R}$ for maps h corresponding to wave maps u which depend continuously on time in the H^1-topology.

Note that formally φ satisfies

$$\varphi_{tt} - \varphi_{rr} - \frac{3}{r}\varphi_r + \frac{f(r\varphi) - r\varphi}{r^3} = 0.$$

Expanding

$$f(r\varphi) = g(r\varphi)g'(r\varphi) = \left(r\varphi + \frac{g'''(0)}{6}(r\varphi)^3 + \dots\right)\left(1 + \frac{g'''(0)}{2}(r\varphi)^2 + \dots\right)$$

$$= r\varphi + \frac{2}{3}g'''(0)(r\varphi)^3 + \dots,$$

we can express the nonlinear term as

$$\frac{f(r\varphi) - r\varphi}{r^3} = -K(r\varphi)\varphi^3,$$

where K is smooth, $K(h) = K(-h)$, and

$$K(0) = -\frac{2}{3}g'''(0)$$

equals $2/3$ the curvature of N at 0.

Also regarding

$$\varphi_{tt} - \varphi_{rr} - \frac{3}{r}\varphi_r = \Box\varphi,$$

as the wave operator on $\mathbb{R} \times \mathbb{R}^4$, acting on the radially symmetric function $\varphi(t,x) = \varphi(t,r)$ for $x \in \mathbb{R}^4$ with $|x| = r$, thus we arrive at the equation

$$\Box\varphi - K(r\varphi)\varphi^3 = 0 \qquad \text{on } \mathbb{R} \times \mathbb{R}^4. \tag{3.9}$$

In case (3.1), moreover, (3.8) translates into the boundary condition

$$\varphi(t,r) = 0 \qquad \text{for } r \geq R_0 + |t|. \tag{3.10}$$

Finally, in case (3.1), (3.2) hold, using boundedness of h (Lemma 3.3) we infer that there exists a constant C, possibly depending on h, such that

$$h(t,r) \leq Cg(h(t,r)).$$

Hence

$$E_{\text{equi}}\big(h(t)\big) \geq \frac{1}{2}\int_0^\infty \left\{ |h_t|^2 + |h_r|^2 + \frac{h^2}{C^2 r^2} \right\} r\, dr$$

$$= \frac{1}{2}\int_0^\infty \left\{ |\varphi_t|^2 + |\varphi_r + \frac{\varphi}{r}|^2 + \frac{\varphi^2}{C^2 r^2} \right\} r^3\, dr$$

and it follows that (φ, φ_t) is bounded in the energy space $H_c^1(\mathbb{R}^4; T\mathbb{R})$.

Thus, solutions h of (3.4) of class H^1 with data $(h_0, h_1) \in H_c^1$ correspond to solutions φ of (3.9) of class H^1 with data in H_c^1, and conversely.

For a compact target surface, with g satisfying (3.3), condition (3.8) translates into

$$\varphi(t, r) = -\frac{d(t)q_1}{r} \qquad \text{for } r \geq R_0 + |t|. \tag{3.11}$$

Remark that $d(0) = d_0 \in \mathbb{Z}$ corresponds to the degree of the initial map

$$u_0: \mathbb{R}^2 \cong S^2 \to N \cong S^2.$$

Hence for initial degree $d_0 \neq 0$, from (3.11) we conclude that

$$\varphi_r(t, r) = \frac{d_0 q_1}{r^2} \notin L^2\big([0, \infty[\big)$$

for small t. H^1-solutions h of (3.4), (3.6) with $d_0 \neq 0$ therefore do *not* correspond to solutions φ of (3.9) of class H^1, but only to solutions which are *locally* of class H^1.

3.2.3.   Well-posedness, a model case.   To set the stage for the general result first consider the model case

$$g^2(h) = h^2 + \frac{1}{2}h^4.$$

Equation (3.9) in this case reads

$$\Box\varphi + \varphi^3 = 0 \qquad \text{on } \mathbb{R} \times \mathbb{R}^4 \tag{3.12}$$

with Cauchy data

$$\varphi\,|_{t=0} = \varphi_0, \qquad \varphi_t\,|_{t=0} = \varphi_1, \tag{3.13}$$

where $(\varphi_0, \varphi_1) \in H_c^1(\mathbb{R}^4; T\mathbb{R})$.

Equation (3.12) is a special case of a class of semi-linear wave equations involving critical growth exponents for which a full theory of existence, uniqueness and regularity has been developed in the past years, starting with the work of Struwe [48] on radial solutions of the equation

$$\Box\varphi + \varphi^5 = 0 \qquad \text{on } \mathbb{R} \times \mathbb{R}^3$$

in 1988. The symmetry condition in [48] was removed by Grillakis [20], still in 3 space dimensions.

The insight how to treat the higher-dimensional cases, in particular the case $m = 4$ which is relevant here, came from the work of Kapitanskii [30] who pointed out the use of the Strichartz inequalities for the analysis of semilinear wave equations like (3.12).

Grillakis [21] then was the first to realize that the Strichartz estimates and the crucial decay estimate from [48] could be combined to prove regularity for the equation

$$\Box \varphi + \varphi |\varphi|^{2^* - 2} = 0 \qquad \text{in } \mathbb{R} \times \mathbb{R}^m \tag{3.14}$$

in dimensions $m \le 5$, where $2^* = \frac{2m}{m-2}$ is the "Sobolev exponent" in dimension m.

More efficient use of the Strichartz estimates was then made by Shatah-Struwe [42], [43] who extended the regularity results for (3.14) to dimensions $m \le 7$ and, moreover, proved that the initial value problem for (3.14) with finite energy data $(\varphi_0, \varphi_1) \in H_c^1$ is well-posed in all dimensions $m \ge 3$.

In particular, the result from [43] applies to the Cauchy problem for (3.12), (3.13).

Theorem 3.4 *For any $(\varphi_0, \varphi_1) \in H_c^1$ there exists a unique solution φ of (3.12), (3.13) such that $(\varphi, \varphi_t) \in C^0(\mathbb{R}; H_c^1) \cap L^q(\mathbb{R}; \dot{B}_q^{1/2} \times \dot{B}_q^{-1/2})$, where $q = \frac{10}{3}$ and $L^q(\mathbb{R}; \dot{B}_q^{1/2})$ is the Besov space of functions φ with "half a spatial derivative" in $L^q(\mathbb{R} \times \mathbb{R}^4)$. If $(\varphi_0, \varphi_1) \in C^\infty$, then $\varphi \in C^\infty$, as well.*

Remark that by Sobolev's embedding $\varphi \in L^\infty(\mathbb{R}; L^4(\mathbb{R}^4)) \cap L^{\frac{10}{3}}(\mathbb{R}; L^{\frac{40}{7}}(\mathbb{R}^4))$ and hence by interpolation

$$\varphi \in L^5(\mathbb{R} \times \mathbb{R}^4).$$

The *proof of uniqueness* now follows easily from the Strichartz estimate

$$\|\psi\|_{L^{\frac{10}{3}}(\mathbb{R} \times \mathbb{R}^4)} \le C \|\Box \psi\|_{L^{\frac{10}{7}}(\mathbb{R} \times \mathbb{R}^4)} \tag{3.15}$$

for $\psi \colon \mathbb{R} \times \mathbb{R}^4$ with vanishing Cauchy data

$$\psi|_{t=0} = 0 = \psi_t|_{t=0}.$$

For simplicity we assume that the initial data (3.13) have small energy. Then the square of the L^5-norm of any solution φ of (3.12), (3.13) as in Theorem 3.4 is bounded by the energy of the initial data. Let $\varphi, \tilde{\varphi}$ be two solutions of (3.12) as in Theorem 3.4 sharing Cauchy data $(\varphi_0, \varphi_1) \in H_c^1$ at $t = 0$. Then $\psi = \varphi - \tilde{\varphi}$ satisfies

$$|\Box \psi| = |\varphi^3 - \tilde{\varphi}^3| \le C |\psi| (\varphi^2 + \tilde{\varphi}^2)$$

and hence by (3.15) and Hölder's inequality

$$\|\psi\|_{L^{\frac{10}{3}}(\mathbb{R} \times \mathbb{R}^4)} \le C \|\psi\|_{L^{\frac{10}{3}}(\mathbb{R} \times \mathbb{R}^4)} \left(\|\varphi\|_{L^5(\mathbb{R} \times \mathbb{R}^4)}^2 + \|\tilde{\varphi}\|_{L^5(\mathbb{R} \times \mathbb{R}^4)}^2 \right)$$
$$\le CE(u(0)) \|\psi\|_{L^{\frac{10}{3}}(\mathbb{R} \times \mathbb{R}^4)},$$

which implies that $\psi = 0$, if $E(u(0))$ is sufficiently small. $\Box$

The slightly improved space-time integrability of the solutions obtained in Theorem 3.4 (that is, $u \in L^5$ instead of L^4, only) also suffices to propagate further regularity of the data; in particular, φ is smooth if the data are smooth; see also [19] for corresponding results in higher dimensions.

The Strichartz inequality can be localized to light cones; moreover, for small energy the estimates relevant for the proof of Theorem 3.4 in the model case (3.12) continue to be valid for the general equation (3.9) with data $(\varphi_0, \varphi_1) \in H_c^1$; see [42]. Hence it seems that this problem and therefore the Cauchy problem for co-rotational wave maps into surfaces of revolution satisfying (3.1), (3.2) is globally well-posed for small initial energy. If the energy is large, it is conceivable that concentration may occur at a finite number of points $(t_i, 0)$, just as in the case of the heat flow for harmonic maps of surfaces; see Theorem 2.3. However, a curvature condition or a "small range" condition can prevent energy from concentrating; see [7], [22], [42], [44].

Under assumption (3.3), energy estimates for φ always need to be localized; moreover, the degree of the corresponding map u may change at concentration points.

We close this section by stating the following result implied by this reasoning.

Theorem 3.5 *i) Suppose N satisfies (3.1), (3.2). Then for any co-rotational data $(h_0, h_1) \in H_c^1$ corresponding to data $(\varphi_0, \varphi_1) \in H_c^1$ for (3.12), there exists a unique global weak solution $h(t, r) = r\varphi(t, r)$ of (3.4), respectively (3.12), in the following class:*
There are numbers $0 = t_0 < t_1 < t_2 < \ldots < t_I < t_{I+1} = \infty$, such that $(\varphi, \varphi_t) \in C^0([t_i, t_{i+1}[; H_c^1) \cap L_{\mathrm{loc}}^q([t_i, t_{i+1}[; \dot{B}_q^{1/2}(\mathbb{R}^4))$ for $0 \le i \le I$, $q = 10/3$, and

$$E_{\mathrm{equi}}(h(t)) \equiv E_i \qquad \text{in } [t_i, t_{i+1}[,$$

where $E_{i+1} \le E_i - \epsilon_0$ for each i for some $\epsilon_0 = \epsilon_0(N) > 0$. In particular, $I \le E_0/\epsilon_0$. (Similarly for negative time.)

ii) If N is compact, the same result holds; however, the estimates for φ are only local in space. At each singularity, the topological degree of the corresponding map u may change.

Details will be given in the forthcoming thesis of Wilhelmy. We conjecture that concentration, as described in Theorem 3.5, can in fact only occur for compact targets as in part ii). Whether concentration of energy ever occurs in finite time, however, still remains as one of the most challenging open problems in this field.

3.3. Towards well-posedness for general targets

Now that we have found confidence that the Cauchy problem for wave maps may, indeed, be well-posed in the energy space for a $(1 + 2)$-dimensional space-time domain, we drop the symmetry assumption on N and the map u and return to the general setting. Thus, for given Cauchy data

$$(u_0, u_1) \in H_c^1(M; TN)$$

we hope to show the existence of a unique, global, weak solution $u\colon \mathbb{R} \times M \to N$ of the Cauchy problem

$$\Box u = -A(u)(\partial_\alpha u, \partial^\alpha u) \perp T_u N, \tag{3.16}$$

$$u\mid_{t=0} = u_0, \ u_t\mid_{t=0} = u_1. \tag{3.17}$$

While we cannot yet solve this problem, we will discuss an approximation method to prove existence and we present some partial results on the relevant convergence problem.

3.4. Approximate solutions

3.4.1. Penalty method. Suppose as in Section 1 that N is compact, isometrically embedded in $\mathbb{R}^d$, and there is a tubular neighborhood $U_{2\delta}(N)$ of width 2δ of N with smooth nearest neighbor projection $\pi_N\colon U_{2\delta}(N) \to N$. Also let $\chi \in C^\infty(\mathbb{R})$ be a function such that $\chi(s) = s$ for $s \le \delta^2$, $\chi(s) \equiv \mathrm{const.}$ for $s \ge 2\delta^2$, $\chi' \ge 0$, $\chi'' \le 0$, and let $\mathrm{dist}(p, N)$ be the distance from a point $p \in \mathbb{R}^d$ to its nearest neighbor on N. Note that

$$p \mapsto \frac{1}{2}\chi\big(\mathrm{dist}^2(p, N)\big)$$

then extends to a smooth function on $\mathbb{R}^d$ whose gradient is given by

$$\chi'\big(\mathrm{dist}^2(p, N)\big)\big(p - \pi_N(p)\big),$$

if $p \in \mathcal{U}_{2\delta}(N)$, and 0 otherwise.

For $L \in \mathbb{N}$ consider solutions $u^L\colon \mathbb{R} \times M \to \mathbb{R}^d$ of the Cauchy problem

$$\Box u^L + L\chi'\big(\mathrm{dist}^2(u^L, N)\big)\big(u^L - \pi_N(u^L)\big) = 0 \qquad \text{on } \mathbb{R} \times M \tag{3.18}$$

with data $(u_0, u_1) \in H_c^1(M; TN)$ at $t = 0$.

Equation (3.18) implies the conservation law

$$\frac{d}{dt}\left(e(u^L) + \frac{L}{2}\chi\big(\mathrm{dist}^2(u^L, N)\big)\right) - \mathrm{div}\langle \nabla u^L, u_t^L\rangle = 0, \tag{3.19}$$

where

$$e(u^L) = \frac{1}{2}|Du^L|^2$$

is the energy density for the free wave equation. Let

$$e_L(u^L) = e(u^L) + \frac{L}{2}\chi\big(\mathrm{dist}^2(u^L, N)\big)$$

and let

$$E_L\big(u^L(t)\big) = \int_M e_L\big(u^L(t)\big)\, dx.$$

Upon integrating (4.4) over M, thus we find that

$$\frac{d}{dt} E_L\left(u^L(t)\right) = 0;$$

in particular,

$$E_L\left(u^L(t)\right) = E_L\left(u^L(0)\right) = E\left(u^L(0)\right) = \frac{1}{2} \int_M \left\{ |u_1|^2 + |\nabla u_0|^2 \right\} dx$$

for all L and all t. It follows that, as $L \to \infty$, a sub-sequence

$$Du^L \to Du \quad \text{weakly-}* \text{ in } L^\infty\left(\mathbb{R}; L^2(M)\right).$$

Moreover, since $u^L(0) = u_0$ for all L, and in view of the compactness of the restriction (trace) operator $H^{1,2}(\mathbb{R} \times M) \hookrightarrow L^2(\{t\} \times M)$ for any t,

$$u^L \to u \quad \text{in } L^2(M), \text{ locally}$$

uniformly in time. Hence, by Fatou's lemma, also

$$\int_{\{t\} \times M} \chi\left(\text{dist}^2(u, N)\right) dx \leq \liminf_{L \to \infty} \int_{\{t\} \times M} \chi\left(\text{dist}^2(u^L, N)\right) dx$$

$$\leq \frac{1}{L} E_L\left(u^L(t)\right) \to 0 \quad (L \to \infty)$$

for any t, and it follows that $u: \mathbb{R} \times M \to N$. Finally, since (3.18) has propagation speed ≤ 1, for data $(u_0, u_1) \in H_c^1$ also $Du^L(t)$ has uniformly compact support for any t and thus

$$(u, u_t) \in L^\infty\left(\mathbb{R}; H_c^1(M; TN)\right).$$

However, while (3.18) implies that

$$\Box u^L \perp T_{\pi_N(u^L)} N$$

at all points in space-time, it is not clear that this relation, and hence (3.16), persists in the limit $L \to \infty$. In special cases, the analysis is, in fact, quite simple.

3.4.2. The sphere. We slightly modify the approximation scheme if $N = S^k$.
    For $L \in \mathbb{N}$, following Shatah [41], we consider solutions $u^L: \mathbb{R} \times M \to \mathbb{R}^d$, $d = k + 1$, of the equation

$$\Box u^L + L\left(|u^L|^2 - 1\right)u^L = 0 \qquad \text{on } \mathbb{R} \times M \tag{3.20}$$

with Cauchy data $(u_0, u_1) \in H_c^1(M; TN)$ at $t = 0$. The initial value problem for (3.20) admits global weak solutions u^L such that

$$\tilde{E}_L\left(u^L(t)\right) = E\left(u^L(t)\right) + \frac{L}{4} \int_M ||u^L|^2 - 1|^2 \, dx = \tilde{E}_L\left(u^L(0)\right)$$

$$= E\left(u^L(0)\right) = \frac{1}{2} \int_M \left\{ |u_1|^2 + |\nabla u_0|^2 \right\} dx,$$

uniformly in t, for all L. A sub-sequence (u^L) hence converges to a limit u in the sense that, as $L \to \infty$,

$$u^L \to u \qquad \text{in } L^2(M) \text{ for all } t,$$

$$Du^L \to Du \qquad \text{weakly-* in } L^\infty(\mathbb{R}; L^2(M)),$$

and for any t, by Fatou's lemma.

$$\int_M \left(|u|^2 - 1\right)^2 dx \le \liminf_{L \to \infty} \int_M \left(|u^L|^2 - 1\right)^2 dx = 0.$$

Hence $u: \mathbb{R} \times M \to S^k$.

In order to pass to the limit in (3.20), observe that the nonlinear term always points in the direction of u^L. Taking the exterior product with u^L, thus from (3.20) we obtain the equation

$$\partial^\alpha(\partial_\alpha u^L \wedge u^L) = \Box u^L \wedge u^L = 0 \qquad (3.21)$$

for all L. In the limit $L \to \infty$, therefore also the equation

$$\partial^\alpha(\partial_\alpha u \wedge u) = \Box u \wedge u = 0 \qquad (3.22)$$

is valid in the sense of distributions, which implies that u weakly solves (3.16).

Moreover, multiplying (3.21), (3.22) with a 2-vector $\varphi \in C_0^\infty(\mathbb{R} \times M)$ and integrating by parts on $[0, \infty[\times M$, we obtain

$$\int_0^\infty \int_M \left((\partial_\alpha u^L \wedge u^L) - (\partial_\alpha u \wedge u)\right)\partial^\alpha \varphi \, dx \, dt$$

$$= \int_{\{0\} \times M} (u_1 \wedge u_0 - u_t \wedge u_0)\varphi \, dx$$

In the limit $L \to \infty$, the left hand side vanishes. Since $\varphi(0, \cdot)$ is arbitrary, thus we conclude that

$$\left(u_t(0) - u_1\right) \wedge u_0 = 0;$$

that is, $u_t(0) = u_1$ in the sense of traces. Here we used the fact that both u_1 and $u_t(0)$ are tangent to S^k along u_0, that is,

$$\langle u_1, u_0 \rangle = \langle u_t(0), u_0 \rangle = 0.$$

Therefore, as $t \to 0$, $u_t(t, \cdot) \to u_1$ weakly in L^2 as $t \to 0$. On the other hand

$$\limsup_{t \to 0} \frac{1}{2} \int_M |u_t(t, \cdot)|^2 \, dx + \frac{1}{2} \int_M |\nabla u_0|^2 \, dx \le \limsup_{t \to 0} E\big(u(t)\big)$$

$$\le \limsup_{t \to 0} \liminf_{L \to \infty} \tilde{E}_L\big(u^L(t)\big)$$

$$\le E\big(u^L(0)\big) = \frac{1}{2} \int_M \left(|u_1|^2 + |\nabla u_0|^2\right) dx.$$

It follows that

$$\limsup_{t \to 0} \|u_t(t, \cdot)\|_{L^2} \leqq \|u_1\|_{L^2}.$$

Together with the fact that $u_t(t, \cdot) \rightharpoonup u_1$ weakly in L^2, this implies strong convergence $u_t(t, \cdot) \to u_1$ in L^2 as $t \to 0$. That is, u attains the prescribed initial data continuously in $H_c^1(M; TN)$. Hence we have proved:

Theorem 3.6 *Suppose $N = S^k$, and let $(u_0, u_1) \in H_c^1(M; TN)$. Then there exists a global weak solution u of (3.16), (3.17) of class H^1.*

Remark that u need not be unique; see Section 2.

The above method can be generalized to the orthogonal group $N = SO(n)$, as was observed by Freire [15]. The approximating equation is

$$\Box u + L\nabla_u F(u) = 0, \tag{3.23}$$

where

$$F(u) = \int_M |u^t u - \mathbf{1}|^2 \, dx,$$

and where u^t denotes the transposed matrix u and $|\cdot|$ is the norm induced by the scalar product

$$(A, B) = \text{trace} \, (A^t B).$$

Given Cauchy data $(u_0, u_1) \in H_c^1(M, TSO(n))$ at $t = 0$, for any L there exists a solution $u^L \colon \mathbb{R} \times M \to \mathbb{R}^{n \times n}$ of (4.8) with

$$u^L \mid_{t=0} = u_0, \quad u_t^L \mid_{t=0} = u_1.$$

Moreover, (u^L, u_t^L) is bounded in $L^\infty\big(\mathbb{R}; H_c^1(M; T\mathbb{R}^{n \times n})\big)$ with $Du^L(t)$ having uniformly compact support and, as $L \to \infty$, a sub-sequence

$$u^L \to u \qquad \text{in } L^2(M), \text{ locally uniformly in } t,$$
$$Du^L \rightharpoonup Du \qquad \text{weakly-} * \text{ in } L^\infty\big(\mathbb{R}; L^2(M)\big),$$

where $u \colon \mathbb{R} \times M \to SO(n)$ and $(u, u_t) \in L^\infty\big(\mathbb{R}; H_c^1(M; TSO(n))\big)$.

Remark that

$$T_u SO(n) = u T_\mathbf{1} SO(n),$$

where $T_\mathbf{1} SO(n) = so(n)$ denotes the Lie algebra of $SO(n)$. Recall that $so(n)$ consists precisely of the anti-symmetric matrices A, $A^t = -A$. Moreover, the orthogonal complement of $T_u SO(n)$ with respect to $(\cdot, \cdot)$,

$$T_u^\perp SO(n) = u T_\mathbf{1}^\perp SO(n),$$

where $T_\mathbf{1}^\perp SO(n) = \big(so(n)\big)^\perp$, consists precisely of the symmetric matrices $B = B^t$.

Indeed, any matrix M may be split

$$M = \frac{1}{2}(M + M^t) + \frac{1}{2}(M - M^t)$$

into its symmetric and anti-symmetric part and we have

$$
\begin{aligned}
(A, B) &= \text{trace}\,(A^t B) = -\text{trace}\,(AB) \\
&= \text{trace}\,(B^t A) = \text{trace}\,(BA) = \text{trace}\,(AB) = 0
\end{aligned}
$$

whenever $A = -A^t$, $B = B^t$.

Similarly, since $F(u) = F(u^t)$, we have

$$u^t \nabla_u F(u) = \big(\nabla_u F(u)\big)^t u$$

for any u. Thus, from (3.23) we obtain

$$\partial^\alpha \big((u^L)^t \partial_\alpha u^L - \partial_\alpha (u^L)^t u^L\big) = (u^L)^t \Box u^L - \Box (u^L)^t u^L = 0$$

for every L. Passing to the limit $L \to \infty$, then we find

$$\partial^\alpha (u^t \partial_\alpha u - \partial_\alpha u^t u) = u^t \Box u - \Box u^t u = 0;$$

that is, $u^t \Box u \in \big(so(u)\big)^\perp$. Since $u \in SO(n)$, we have $uu^t = \mathbf{1}$ and thus, finally,

$$\Box u \in u\big(so(u)\big)^\perp = T_u^\perp SO(u),$$

as desired.

It is conceivable that Theorem 3.6 extends to any homogeneous space as target.

Observe that the condition $u_1 \in T_{u_0} N$ is crucial in showing that the initial data are attained continuously in L^2. If $u_1 \notin T_{u_0} N$ one can show that as $t \to 0$ we have convergence $u_t(t) \to d\pi_N(u_0)u_1$, the projection of u_1 to $T_{u_0} N$, weakly in L^2. However, due to the loss of the energy of the normal component, we cannot show strong convergence.

Added in proof By recent work of Freire (final version of [15]) and Yi Zhou ([52]), Theorem 3.6 is indeed true for any compact homogeneous space N.

3.5. Convergence
For general targets the problem of convergence is more difficult. Let us first consider the stationary case.

3.5.1. The stationary case. Suppose, for simplicity, that M is a compact surface without boundary; for instance, $M = T^2 = \mathbb{R}^2/\mathbb{Z}^2$. Moreover, let N be a compact k-dimensional manifold, without boundary, isometrically embedded into some Euclidean $\mathbb{R}^d$, and suppose, for simplicity, that TN admits a smooth orthonormal frame field $(\bar{e}_1, \ldots, \bar{e}_k)$. That is, at each point $p \in N$ the collection $(\bar{e}_1(p), \ldots, \bar{e}_k(p))$ is an orthonormal basis for $T_p N$, smoothly varying with p.

By a construction due to Hélein [25] and Christodoulou-Tahvildar-Zadeh [7] this latter hypothesis can be made without loss of generality in the context of harmonic maps. Indeed, if the original target N does not have a parallelizable tangent bundle we can embed N as a totally geodesic submanifold of a compact manifold $\tilde{N}$ that has this property by taking two copies of a tubular neighborhood of N in $\mathbb{R}^d$, endowed with the product metric of $N \times \mathbb{R}^{d-k}$, and gluing them together along their boundaries. The standard basis of $\mathbb{R}^d$ then yields the desired frame field for $T\tilde{N}$, at least near the range N of our maps. Moreover, since $N \subset \tilde{N}$ is totally geodesic, for any map $u \colon M \to N \subset \tilde{N} \subset \mathbb{R}^{\tilde{d}}$ the component orthogonal to $T_u N$ of the Laplacian Δu in $T_u \tilde{N}$ vanishes. In particular, a harmonic map $u \colon M \to N$ will be harmonic, regarded as a map $u \colon M \to \tilde{N}$. Henceforth, therefore we replace N by $\tilde{N}$ and assume $N = \tilde{N}$.

Consider a sequence (u^L) of maps $u^L \in H^{1,2}(M; N)$ such that $u^L \rightharpoonup u$ weakly in H^1 and suppose

$$\Delta u^L + f^L \perp T_{u^L} N, \tag{3.24}$$

where

$$f^L \to 0 \qquad \text{strongly in } H^{-1},$$

the dual of $H^{1,2}(M; \mathbb{R}^d)$.

Theorem 3.7 *Under the above assumptions, u is (weakly) harmonic.*

This result is due to Bethuel [1]. A drastically simplified proof was recently given by Freire-Müller-Struwe [16], which we present below.

From now on, moreover, it is convenient to use the language of differential forms. Thus, we let d, δ be the exterior differential and co-differential, respectively. For a 1-form $\varphi = \varphi_\alpha \, dx^\alpha$ we have $\delta\varphi = \frac{\partial}{\partial x^1}\varphi_1 + \frac{\partial}{\partial x^2}\varphi_2$, for a 2-form $b = \beta \, dx^1 \wedge dx^2$ we have $\delta b = -\frac{\partial}{\partial x^2}\beta \, dx^1 + \frac{\partial}{\partial x^1}\beta \, dx^2$. Moreover, we define the Hodge Laplacian on forms as $\Delta = d\delta + \delta d$, acting as the standard Laplacian on the coefficients of the forms. (We always assume $M = T^2 = \mathbb{R}^2/\mathbb{Z}^2$, so that we can use the standard 1-forms dx^1, dx^2 as basis.) Finally, we contract 1-forms $\varphi = \varphi_\alpha \, dx^\alpha, \psi = \psi_\alpha \, dx^\alpha$ using the metric on $M = T^2$ by letting $\varphi \cdot \psi = \varphi_1\psi_1 + \varphi_2\psi_2$.

Let $\bar{e}_1, \ldots, \bar{e}_k$ denote the orthonormal frame for TN. Then for each L the collection $\bar{e}_1 \circ u^L, \ldots, \bar{e}_k \circ u^L$ is a frame for the pulled back bundle $(u^L)^{-1}TN$; that is, at each $x \in M$, the collection $\bar{e}_1(u^L(x)), \ldots, \bar{e}_k(u^L(x))$ is an orthonormal base for $T_{u^L(x)}N$. Other such frames may be obtained by rotating this frame; that is, by letting

$$e_i^L(x) = R_{ij}^L(x)\bar{e}_j(u^L(x)), \qquad \text{where } R^L = (R_{ij}^L) \in SO(k).$$

We express du (respectively du^L) in terms of e_i (respectively e_i^L) as

$$du = \theta_i e_i, \quad \theta_i = \langle du, e_i \rangle.$$

Also denote the connection 1-form of a frame field (e_i) as

$$\omega_{ij} = \langle de_i, e_j \rangle.$$

Note that

$$\langle \Delta u, e_i \rangle = \delta \theta_i - \omega_{ij} \cdot \theta_j.$$

Hence u is harmonic if and only if

$$\delta \theta_i = \omega_{ij} \cdot \theta_j \qquad \text{for any } i \tag{3.25}$$

in the distribution sense.

Note that the frames $(e_i), (e_i^L)$ are only determined up to rotation, that is, up to a gauge transformation in the bundle of frames. In particular, by choosing $e_i = R_{ij}(\bar{e}_j \circ u)$ such that (e_i) minimizes

$$\Sigma_i \int_M |\nabla e_i|^2 \, dx,$$

we obtain the following result of Hélein [25].

Lemma 3.8 *For any $u \in H^{1,2}(M; N)$ there exists a frame (e_i) for $u^{-1}TN$ such that the associated connection satisfies the Coulomb gauge condition*

$$\delta \omega_{ij} = 0, \quad 1 \le i, j \le k.$$

Moreover,

$$\Sigma_i \int_M |\nabla e_i|^2 \, dx \le \Sigma_i \int_M |\nabla (\bar{e}_i \circ u)|^2 \, dx \le CE(u).$$

In the following we assume that $(e_i), (e_i^L)$ are in Coulomb gauge. Hence, in particular, (e_i^L) is bounded in H^1 and, passing to a further sub-sequence, if necessary, we may assume that $e_i^L \rightharpoonup e_i$ weakly in H^1 and $\omega_{ij}^L \rightharpoonup \omega_{ij}$ weakly in L^2, as $L \to \infty$. By Hodge decomposition, for ω_{ij} (respectively, ω_{ij}^L) we have

$$\omega_{ij} = da_{ij} + \delta b_{ij} + H_{ij},$$

where a_{ij} and b_{ij} are normalized by the condition

$$\int_M a_{ij} \, dx = \int_M b_{ij} = 0,$$

and where H_{ij} is a harmonic 1-form (a constant linear combination of dx^1, dx^2 if $M = T^2$). By mutual orthogonality,

$$\|da_{ij}\|_{L^2}^2 + \|\delta b_{ij}\|_{L^2}^2 + \|H_{ij}\|_{L^2}^2 = \|\omega_{ij}\|_{L^2}^2.$$

In particular (b_{ij}^L) is bounded in H^1, (H_{ij}^L) is bounded in any smooth topology, and we may assume that $b_{ij}^L \to b_{ij}$ weakly in H^1 as $L \to \infty$, while $H_{ij}^L \to H_{ij}$ smoothly.

Moreover, the Coulomb gauge condition implies

$$\delta\omega_{ij} = \Delta a_{ij} = 0;$$

hence $a_{ij} = 0$, and similarly for a_{ij}^L.

Consider the term

$$\delta b_{ij}^L \cdot \theta_j^L - \delta b_{ij} \cdot \theta_j.$$

Let us fix, say, $i = 1$ and consider only the term involving $j = 2$ in this sum. For brevity we write θ, θ^L instead of θ_1, θ_1^L, etc. Let

$$b = \beta \, dx^1 \wedge dx^2, \quad b^L = \beta^L \, dx^1 \wedge dx^2.$$

Then

$$\delta b^L \cdot \theta^L = \langle \frac{\partial}{\partial x^1}\beta^L \frac{\partial}{\partial x^2}u^L - \frac{\partial}{\partial x^2}\beta^L \frac{\partial}{\partial x^1}u^L, e_2^L \rangle$$

has the structure of a Jacobian determinant. Due to this particular structure, a special weak compactness property holds. In fact, from [35], Lemma IV. 3 we have the following lemma.

Lemma 3.9

$$\delta b^L \cdot \theta^L \to \delta b \cdot \theta + \Sigma_{j \in J}\nu_j \delta_{x_j}$$

weakly in the sense of distributions, where J is an at most countable set.

Since J is countable, the capacity of the set $X = \{x_j\}_{j \in J}$ vanishes and there exists a sequence of functions $\varphi_l \in C^\infty(M)$, $0 \le \varphi_l \le 1$, such that $\varphi_l \equiv 0$ in a neighborhood of X for each l and $\varphi_l \to 1$ in $H^{1,2}(M)$ as $l \to \infty$.

Hence for any $\varphi \in C^\infty$ we have

$$\int_M (\delta\theta_i - \omega_{ij} \cdot \theta_j)\varphi \, dx = \lim_{l \to \infty} \int_M (\delta\theta_i - \omega_{ij} \cdot \theta_j)\varphi\varphi_l \, dx$$

and for the proof of (3.25) it suffices to show that

$$\int_M (\delta\theta_i - \omega_{ij} \cdot \theta_j)\varphi \, dx = 0 \qquad (3.26)$$

for any $\varphi \in C^\infty$ vanishing near X.

Now we use our assumption that

$$\delta\theta_i^L - \omega_{ij}^L \cdot \theta_j^L = \langle \Delta u^L, e_i^L \rangle = \langle f^L, e_i^L \rangle \to 0$$

as $L \to \infty$ in the sense of distributions.

Moreover, since $\theta_i^L \rightharpoonup \theta_i$ weakly in L^2, we also have weak convergence $\delta\theta_i^L \rightharpoonup \delta\theta_i$ in the distribution sense. Thus, with error terms $o(1) \to 0$ as $L \to \infty$, we have

$$\int_M (\delta\theta_i - \omega_{ij} \cdot \theta_j)\varphi\, dx = \int_M (\delta\theta_i^L - \omega_{ij} \cdot \theta_j)\varphi\, dx + o(1)$$

$$= \int_M (\omega_{ij}^L \cdot \theta_j^L - \omega_{ij} \cdot \theta_j)\varphi\, dx + o(1)$$

$$= \int_M (\delta b_{ij}^L \cdot \theta_j^L - \delta b_{ij} \cdot \theta_j)\varphi\, dx + o(1)$$

and the latter tends to 0 as $L \to \infty$ by Lemma 3.9, on account of the fact that φ is supported away from X. This proves (3.26) and the Theorem.

3.5.2.   The time-dependent case.   We consider the following model situation. Let (u^L) be a sequence of wave maps $u^L \colon \mathbb{R} \times M \to N$ with $E(u^L(t)) \le E(u^L(0)) \le C$, uniformly in t and L. We may assume that, as $L \to \infty$,

$$u^L \to u \qquad \text{in } L^2(M), \text{ locally uniformly in time}$$
$$Du^L \rightharpoonup Du \qquad \text{weakly-$*$ in } L^\infty\big(\mathbb{R}; L^2(M)\big).$$

Then, by a result of Freire-Müller-Struwe [16] there holds

Theorem 3.10 *Under the above assumptions, the limit map $u \colon \mathbb{R} \times M \to N$ weakly solves the wave map equation (3.16).*

Below, we indicate the main steps in the proof of Theorem 3.10.

Again, as in the stationary case, we may assume that TN is parallelizable. Let $\bar{e}_1, \dots, \bar{e}_k$ be an orthonormal frame field for TN and for u, respectively u^L, consider corresponding rotated frames for the pull-back bundle $u^{-1}TN$, given by

$$e_i(z) = R_{ij}(z)\bar{e}_j\big(u(z)\big) \qquad \text{for } z = (t, x) \in \mathbb{R} \times M,$$

where

$$R = (R_{ij}) \colon \mathbb{R} \times M \to SO(k).$$

Note that (3.16) is equivalent to the relation

$$\langle \Box u, e_i \rangle = \partial^\alpha \theta_{i,\alpha} - \omega_{ij}^\alpha \theta_{j,\alpha} = 0$$

in the distribution sense. (Recall that we raise indeces with the Minkowski metric.)

That is, for the proof of Theorem 3.10 we have to show that for any $\varphi \in C_0^\infty$ and any $\epsilon > 0$ there holds

$$\left| \int_{\mathbb{R} \times M} (\partial^\alpha \theta_{i,\alpha} - \omega_{ij}^\alpha \theta_{j,\alpha})\varphi\, dz \right| < \epsilon. \tag{3.27}$$

Fix such φ and $\epsilon > 0$. The energy inequality for (3.16) implies:

Lemma 3.11 *There is a sub-sequence* (u^L) *such that the ϵ-concentration set of* (u^L),

$$S_\epsilon = \left\{ z_0 = (t_0, x_0); \forall R > 0 : \limsup_{L \to \infty} \int_{B_R(x_0; \mathbb{R}^2)} |Du^L(t_0)|^2 \, dx \geq \epsilon \right\}$$

has vanishing $H^{1,2}$-capacity; that is, there exists a sequence of cut-off functions $\varphi_l \in H^{1,2} \cap L^\infty(\mathbb{R} \times M)$ such that $0 \leq \varphi_l \leq 1$, $\varphi_l \equiv 0$ in a neighborhood of S_ϵ and $\varphi_l \to 1$ in $H^{1,2}$ as $l \to \infty$.

Since

$$\int_{\mathbb{R} \times M} (\partial^\alpha \theta_{i,\alpha} - \omega_{ij}^\alpha \theta_{j,\alpha}) \varphi_l \varphi \, dz \to \int_{\mathbb{R} \times M} (\partial^\alpha \theta_{i,\alpha} - \omega_{ij}^\alpha \theta_{j,\alpha}) \varphi \, dz$$

as $l \to \infty$, it hence suffices to prove (3.27) for testing functions φ that vanish in a neighborhood of S_ϵ. Scaling suitably, we may assume that the support of φ is contained in a fundamental domain Q for $T^3 = \mathbb{R}^3/\mathbb{Z}^3$. Extending u^L, e_i^L, etc. suitably outside Q, we may also regard u^L, e_i^L, etc. as functions on T^3. (The modified functions u^L, of course, only satisfy (3.16) in Q.) On T^3 we impose the Coulomb gauge condition (with respect to the *Euclidean* background metric) by choosing $R^L : T^3 \to SO(k)$ such that

$$\Sigma_i \int_{T^3} |De_i^L|^2 \, dz = \min_R \Sigma_i \int_{T^3} |D(R_{ij}(\bar{e}_j \circ u^L))|^2 \, dz.$$

In this gauge, we have

$$\partial_\alpha \omega_{ij,\alpha} = \delta_{eucl} \omega_{ij} = 0$$

and (e_i^L) is bounded in $H^{1,2}(T^3)$ with

$$\Sigma_i \int_Q |De_i^L|^2 \, dz \leq \Sigma_i \int_Q |D(\bar{e}_i \circ u^L)|^2 \, dz \leq CE(u^L(0)) \leq C.$$

Hence we may assume that $e_i^L \to e_i$ weakly in $H^{1,2}_{loc}(\mathbb{R} \times M)$ and

$$\theta_i^L = \langle du^L, e_i^L \rangle = \theta_{i,\alpha}^L \, dx^\alpha \to \theta_i = \langle du, e_i \rangle,$$
$$\omega_{ij}^L = \langle de_i^L, e_j^L \rangle = \omega_{ij,\alpha}^L \, dx^\alpha \to \omega_{ij} = \langle de_i, e_j \rangle$$

weakly in L^2 as $L \to \infty$.

Moreover, by a simple measure-theoretic argument, the set of concentration points z_0 of (e_i), satisfying

$$\limsup_{r \to 0} r^{-1} \int_{B_r(z_0)} \Sigma_i |De_i|^2 \, dz > 0,$$

has vanishing $H^{1,2}$-capacity and we also may assume that φ vanishes near such points.

Finally, since u^L solves (4.1) and since

$$\partial^\alpha \theta^L_{i,\alpha} \to \partial^\alpha \theta_{i,\alpha} \qquad (L \to \infty)$$

in the distribution sense, for the proof of Theorem 3.10 it suffices to show:

Lemma 3.12 *For $L \in \mathbb{N}$ sufficiently large there holds*

$$\left| \int_{\mathbb{R} \times M} (\omega^{L,\alpha}_{ij} \theta^L_{j,\alpha} - \omega^\alpha_{ij} \theta_{j,\alpha}) \varphi \, dz \right| \leq \epsilon.$$

Proof. By weak convergence $\omega^L_{ij} \rightharpoonup \omega_{ij}$ in $L^2(Q)$,

$$\liminf_{L \to \infty} \int_{T^3} (\omega^{L,\alpha}_{ij} \theta^L_{j,\alpha} - \omega^\alpha_{ij} \theta_{j,\alpha}) \varphi \, dz = \liminf_{L \to \infty} \int_{T^3} \omega^{L,\alpha}_{ij} (\theta^L_{j,\alpha} - \theta_{j,\alpha}) \varphi \, dz$$

In the following, again we consider some fixed pair of indices i, j and we omit these indices for brevity.

Next, let

$$\omega^L = da^L + \delta_{eucl} b^L + H^L$$

be the Hodge decomposition of $\omega^L (= \omega^L_{12})$, normalized by the requirement that

$$\int_{T^3} a^L \, dz = \int_{T^3} b^L \wedge dx^\alpha = 0, \qquad \alpha = 0, 1, 2.$$

By mutual orthogonality

$$\|\delta_{eucl} b^L\|^2_{L^2} + \|H^L\|^2_{L^2} \leq \|\omega^L\|^2_{L^2} \leq C.$$

It follows that $H^L \to H$ smoothly as $L \to \infty$.

Moreover, the Coulomb gauge condition implies $a^L = 0$. Finally,

$$\Delta b^L = d\omega^L = de^L_1 \wedge de^L_2$$

exhibits the crucial determinant structure. $\qquad\qquad\qquad\qquad\qquad \Box$

In the time-dependent setting we will need the following result on Jacobian determinants, due to Coifman-Lions-Meyer-Semmes [8].

Lemma 3.13 *If $\varphi, \psi \in H^{1,2}$ then $d\varphi \wedge d\psi$ belongs to the Hardy space $\mathcal{H}^1$ and $\|d\varphi \wedge d\psi\|_{\mathcal{H}^1} \leq C \|d\varphi\|_{L^2} \|d\psi\|_{L^2}$.*

Decompose

$$\begin{aligned}
(\theta^L - \theta)\varphi &= \langle d((u^L - u)\varphi), e^L \rangle + o(1) \\
&= \langle d(u^L - u)\varphi, e^L - e \rangle + d\langle (u^L - u)\varphi, e \rangle + o(1) \\
&= A^L_1 + A^L_2 + o(1),
\end{aligned}$$

where $o(1) \to 0$ in L^2 as $L \to \infty$.

Denote by $a_i^L, i = 1, 2$, the solution to

$$\Delta a_i^L = \delta_{eucl}^* A_i^L,$$

where δ_{eucl}^* is the adjoint of δ_{eucl} with respect to the Minkowski metric. By using a result of Campanato and Giaquinta [17], we show that the functions $(a_{1,2}^L)$ are bounded in BMO(T^3). In fact, using precise estimates in Morrey spaces and our definition of S_ϵ, we can show that

Lemma 3.14

$$\limsup_{L \to \infty} \|a_1^L\|_{\mathrm{BMO}} \le C\epsilon,$$

$$\limsup_{L \to \infty} \|a_2^L\|_{\mathrm{BMO}} \le C\sqrt{\epsilon}.$$

By $\mathcal{H}^1$-BMO duality (Fefferman-Stein [12]) then we have, with error $o(1) \to 0$ as $L \to \infty$,

$$
\begin{aligned}
\int_{T^3} w^L \cdot (\theta^L - \theta)\varphi\, dz &= \int_{T^3} \delta b^L \cdot (A_1^L + A_2^L)dz + o(1) \\
&= \int_{T^3} b^L \cdot \Delta(a_1^L + a_2^L)dz + o(1) \\
&= \int_{T^3} de_1^L \wedge de_2^L \cdot (a_1^L + a_2^L)dz + o(1) \\
&\le C\|de_1^L \wedge de_2^L\|_{\mathcal{H}^1}\left(\|a_1^L\|_{\mathrm{BMO}} + \|a_2^L\|_{\mathrm{BMO}}\right) + o(1) \\
&\le C\sqrt{\epsilon} + o(1),
\end{aligned}
$$

as desired.

Concluding remark: Observe that the proof of Theorem 3.10 might be adapted to show that the weak H^1-limit of approximate solutions u^L to (3.16) with range on N is a wave map. However, for the sequence (u^L) defined by the penalty method in Section 3.4.1 it is difficult to control the energy of the functions u^L in direction normal to N. For this reason we cannot (yet) use Theorem 3.10 to obtain existence of global weak solutions.

Added in proof The existence of global weak solutions to the Cauchy problem for wave maps $u: \mathbb{R} \times \mathbb{R}^2 \to N$ with initial data of class H^1 was recently obtained by Müller-Struwe [51], using the convergence result outlined in Section 3.5.2 and the viscosity approximation to (1.1) suggested by Yi Zhou [52].

References

[1] F. Bethuel: *Weak convergence of Palais-Smale sequences for some critical functionals*, Preprint (1992)

[2] F. Bethuel: *On the singular set of stationary harmonic maps*, Preprint (1992)

[3] H. Brezis, J.-M. Coron, E. Lieb: *Harmonic maps with defects*, Comm. Math. Phys. **107** (1986) 649–705

[4] K.-C. Chang: *Heat flow and boundary value problem for harmonic maps*, Ann. Inst. H. Poincaré, Analyse Non-Linéaire **6** (1989) 363–395

[5] K.-C. Chang, W.-Y. Ding, R. Ye: *Finite-time blow-up of the heat flow of harmonic maps from surfaces*, J. Diff. Geom. 36 (1992), 507–515

[6] Y. Chen, M. Struwe: *Existence and partial regularity results for the heat flow for harmonic maps*, Math. Z. **201** (1989) 83–103

[7] D. Christodoulou, A. Shadi Tahvildar-Zadeh: *On the regularity of spherically symmetric wave maps*, Preprint

[8] R. Coifman, P.L. Lions, Y. Meyer, S. Semmes: *Compensated compactness and Hardy spaces*, J. Math. Pures Appl. **72** (1993), 247–286

[9] J.-M. Coron: *Nonuniqueness for the heat flow of harmonic maps*, Ann. Inst. H. Poincaré, Analyse Non Linéaire **7** (1990) 335–344

[10] J. Eells, J.H. Sampson: *Harmonic mappings of Riemannian manifolds*, Am. J. Math. **86** (1964) 109–169

[11] L.C. Evans: *Partial regularity for stationary harmonic maps into spheres*, Arch. Rat. Mech. Anal., **116** (1991), 101–163

[12] C. Fefferman, E.M. Stein: H^p *spaces of several variables*, Acta Math. **129** (1972), 137–193

[13] A. Freire: *Uniqueness for the harmonic map flow in two dimensions*, Calc. Var. **3** (1995), 95–105

[14] A. Freire: *Uniqueness for the harmonic map flow from surfaces to general targets*, Comm. Math. Helv. **70** (1995), 310–338, correction Comm. Math. Helv. **71** (1996) 330–337

[15] A. Freire: *Global weak solutions of the wave map system to compact homogeneous spaces*, submitted (1996)

[16] A. Freire, S. Müller, M. Struwe: *Weak convergence of harmonic maps from* $(2 + 1)$-*dimensional Minkowski space to Riemannian manifolds*, Preprint

[17] M. Giaquinta: *Introduction to regularity theory for nonlinear elliptic systems*, Lectures in Mathematics, Birkhäuser Verlag, (1993)

[18] M. Giaquinta, E. Giusti: *The singular set of the minima of certain quadratic functionals*, Ann. Scuola Norm. Sup. Pisa (4) 11 (1984) 45–55

[19] J. Ginibre, A. Soffer, G. Velo: *The global Cauchy problem for the critical non-linear wave-equation*, J. Funct. Anal. **110** (1992), 96–130

[20] M. Grillakis: *Regularity and asymptotic behaviour of the wave equation with a critical nonlinearity*, Ann. of Math. **132** (1990) 485–509

[21] M. Grillakis: *Regularity for the wave equation with a critical nonlinearity*, Comm. Pure Appl. Math. **45** (1992), 749–774

[22] M. Grillakis: *Classical solutions for the equivariant wave map in* $1 + 2$ *dimensions*, Preprint

[23] P. Hartman, A. Wintner: *On the local behavior of solutions of non-parabolic partial differential equations*, Amer. J. Math. **75** (1953), 449–476

[24] F. Hélein: *Regularité des applications faiblement harmoniques entre une surface et une varitée Riemannienne*, C.R. Acad. Sci. Paris Ser. I Math. **312** (1991) 591–596

[25] F. Hélein: *Regularité des applications faiblement harmoniques entre une surface et une varitée Riemannienne*, C.R. Acad. Sci. Paris Ser. I Math. **312** (1991) 591–596

[26] S. Hildebrandt: *Nonlinear elliptic systems and harmonic mappings*, Proc. Beijing Sympos. Diff. Geom. Diff. Eq. 1980, Gordon and Breach (1983) 481–615

[27] L. Hörmander: *Non-linear hyperbolic differential equations*, Lectures 1986–1987, Lund, Sweden

[28] J. Jost: *Harmonic mappings between Riemannian manifolds*, ANU-Press, Canberra, 1984

[29] J. Jost: *Nonlinear methods in Riemannian and Kählerian geometry*, DMV Seminar 10, Birkhäuser, Basel (1991)

[30] L. Kapitanskii: *The Cauchy problem for semilinear wave equations*, part I: J. Soviet Math. **49** (1990), 1166–1186; part II: J. Soviet Math. **62** (1992), 2746–2777; part III: J. Soviet Math. **62** (1992), 2619–2645

[31] S. Klainerman (lecture notes in perparation)

[32] S. Klainerman, M. Machedon: *Space-time estimates for null forms and the local existence theorem*, Preprint

[33] L. Lemaire: *Applications harmoniques de surfaces riemanniennes*, J. Diff. Geom. **13** (1978) 51–78

[34] F.H. Lin: *Une remarque sur l'application $x/|x|$*, C.R. Acad. Sc. Paris **305** (1987) 529–531

[35] P.L. Lions: *The concentration-compactness principle, the limit case II*

[36] C.B. Morrey: *Multiple integrals in the calculus of variations*, Grundlehren 130, Springer, Berlin 1966

[37] T. Rivière: *Applications harmoniques de B^3 dans S^2 partout discounes*, C. R. Acad. Sci. Paris **314** (1992) 719–723

[38] T. Rivière: *Regularité partielle des solutions faibles du problème d'évolution des applications harmoniques en dimension deux*, Preprint

[39] R.S. Schoen, K. Uhlenbeck: *A regularity theory for harmonic maps*, J. Diff Geom. **17** (1982) 307–335, **18** (1983) 329

[40] R.S. Schoen, K. Uhlenbeck: *Boundary regularity and the Dirichlet problem for harmonic maps*, J. Diff. Geom. **18** (1983) 253–268

[41] J. Shatah: *Weak solutions and development of singularities in the $SU(2)$ σ-model*, Comm. Pure Appl. Math. **41** (1988) 459–469

[42] J. Shatah, M. Struwe: *Regularity results for nonlinear wave equations*, Annals of Math. **138** (1993) 503–518

[43] J. Shatah, M. Struwe: *Well-posedness in the energy space for semilinear wave equations with critical growth*, Inter. Math. Res. Notices **7** (1994), 303–309

[44] J. Shatah, A. Tahvildar-Zadeh: *Regularity of harmonic maps from the Minkowski space into rotationally symmetric manifolds*, Comm. Pure Appl. Math. **45** (1992), 947–971

[45] J. Shatah, A. Tahvildar-Zadeh: *Non uniqueness and development of singularities for harmonic maps of the Minkowski space*, Preprint

[46] M. Struwe: *On the evolution of harmonic maps in higher dimensions*, J. Diff. Geom. **28** (1988) 485–502

[47] M. Struwe: *On the evolution of harmonic maps of Riemannian surfaces*, Math. Helv. **60** (1985) 558–581

[48] M. Struwe: *Globally regular solutions to the u^5 Klein-Gordon equation*, Ann. Scuola Norm. Pisa **15** (1988) 495–513

[49] M. Struwe: *Geometric Evolution Problems*, Nonlinear Partial Differential Equations in Differential Geometry, IAS/Park City Mathematics Series vol.2 (AMS, November 1995)

[50] H.C. Wente: *The differential equation $\Delta x = 2H x_u \wedge x_v$ with vanishing boundary values*. Proc. AMS **50** (1975) 59–77

[51] S. Müller, M. Struwe: *Global existence of wave maps in $1+2$ dimensions with finite energy data*, Top. Meth. Nonlin. Analysis, special volume in honor of L. Nirenberg's 70th birthday (to appear)

[52] Yi Zhou: *Global weak solutions for the $1+2$ dimensional wave maps into homogeneous space*, submitted (1996)

Index

BIRKHÄUSER

Progress in Nonlinear Differential Equations and Their Applications

Editor
Haim Brezis,
Département de Mathématiques, Université P. et M. Curie 4, Place Jussieu, 75252 Paris Cedex 05, France and Department of Mathematics, Rutgers University New Brunswick, NJ 08903, U.S.A.

Progress in Nonlinear and Differential Equations and Their Applications is a book series that lies at the interface of pure and applied mathematics. Many differential equations are motivated by problems arising in such diversified fields as Mechanics, Physics, Differential Geometry, Engineering, Control Theory, Biology, and Economics. This series is open to both the theoretical and applied aspects, hopefully stimulating a fruitful interaction between the two sides. It will publish monographs, polished notes arising from lectures and seminars, graduate level texts, and proceedings of focused and refereed conferences.
We encourage preparation of manuscripts in some form of TeX for delivery in camera-ready copy, which leads to rapid publication, or in electronic form for interfacing with laser printers or typesetters.

Proposals should be sent directly to the editor or to: Birkhäuser Boston, 675 Massachusetts Avenue, Cambridge, MA 02139